§sas. | SAS Publishing

MW00905363

Getting Started with the
SAS® Power and Sample Size Application

The Power to Know.

Contents

Acknowledgments

Credits

Documentation

Writing	Wayne Watson
Editing	Virginia Clark, Maura Stokes
Production Support	Tim Arnold, Kirsten Doehler
Cover Design	Creative Solutions Division

Software

Design	Todd Barlow, John Castelloe, Julie LaBarr, Jeff Sun, Wayne Watson
Development	Jeff Sun, Wayne Watson
Installation	Merri Jensen, Susanna Wallenberger
Testing	Sumita Biswas, Kirsten Doehler, Chris Schroter, Jeff Sun, Mark Traccarella, F. Gabriel Younan

Support Group

Technical Support   Craig DeVault

About SAS/STAT Power and Sample Size

SAS/STAT Power and Sample Size (PSS) is a Web-based interface to power and sample size determination analyses in SAS 9.1. The application is accessed through a Web browser (Microsoft Internet Explorer 5.0 or greater).

Documentation

This book describes the features of the SAS/STAT Power and Sample Size application (PSS) and how to use it to perform typical analyses, but it is not intended to teach or describe the statistical methodology that is employed. You can find a description of the statistical techniques used here in the *SAS/STAT User's Guide*, in the chapters for the POWER and GLMPOWER procedures.

Software Requirements

PSS is installed separately from the SAS/STAT product. It is included on the SAS Mid-Tier Components CD. See the installation instructions for the Mid-Tier Components CD in the installation package.

PSS is available in SAS 9.1 for the following platforms: Microsoft Windows NT, 2000, and XP; Sun Solaris 64-bit enabled; HP-UX IPF; and IBM 64-bit enabled AIX.

Web server software is required that supports Servlet/JSP specification 2.3/1.2 and includes a JAXP 1.4-compatible XML parser. PSS is accessed using the Microsoft Internet Explorer browser.

Two configurations are available during installation. With the local configuration, the browser, Web server, and SAS 9.1 must reside on the same machine. With the remote configuration, the browser, Web server, and SAS 9.1 can reside on different machines, and multiple users can access the application on a single machine through the company's intranet.

For both configurations, Base SAS, SAS/STAT, and SAS/GRAPH software must be installed.

For the remote configuration, SAS/CONNECT software must also be installed.

Using This Book

The following font conventions are used throughout this book:

Text Type	Element	Example
bold text	labels	**Description**
	buttons	the **OK** button
	selectable hypertext links	the **Enter means** link
roman text	page names	the Projects page
`monospace text`	check box and radio box choices	`A choice`
	text entered by the user	`Experimental research`

Chapter 1
Overview

Chapter Contents

Chapter 1
Overview

Introduction

When you are planning a study or an experiment, you often need to know how many units to sample to obtain a certain power, or you may want to know the power you would obtain with a specific sample size. The power of a hypothesis test is the probability of rejecting the null hypothesis when the alternative hypothesis is true. With an inadequate sample size, you may not reach valid conclusions with your work; with an excessive sample size, you may waste valuable resources. Thus, performing sample size and power computations is often quite important.

The power and sample size calculations depend on the planned data analysis strategy. That is, if the primary hypothesis test is a two-sample *t* test, then the power calculations must be based on that test. Otherwise, if the sample size calculations and data analyses are not aligned, the results may not be correct.

The SAS/STAT Power and Sample Size application (PSS) is a data analysis tool that provides easy access to power analysis and sample size determination techniques. The application is intended for students and researchers as well as experienced SAS users and statisticians.

Determining sample size requirements ahead of the study is a prospective exercise. You then proceed to select the appropriate number of sampling units and perform data collection and analysis. However, power and sample size calculations are also useful retrospectively. Only prospective power analysis is offered by PSS.

Power and sample size calculations are a function of the specific alternative hypothesis of interest, in addition to other parameters. That is, the power results will vary depending on which value of the alternative hypothesis you specify, so sometimes it is useful to do these analyses for a range of values to determine how sensitive the power analysis is to changes in the alternative hypothesis value. Often, you produce plots of power versus sample size, called power curves, to see how sample size and power affect each other.

The PSS Application

The PSS application is a Web browser application. It is accessed either from your own machine or your organization's intranet using the Microsoft Internet Explorer browser. It relies on the SAS/STAT procedures POWER and GLMPOWER for its computations.

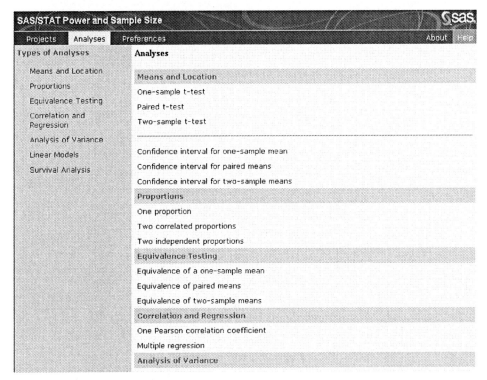

Figure 1.1. PSS Application

This section describes the statistical tasks that are available with the application as well as its principal features.

Analyses

PSS provides power and sample size computations for a variety of statistical analyses. Included are *t* tests, equivalence tests, and confidence intervals for means; exact binomial, chi-square, Fisher's exact, and McNemar tests for proportions; multiple regression and correlation; one-way analysis of variance; linear models, and rank tests for comparing survival curves.

Table 1.1 lists the analyses that are available.

Table 1.1. Available Analyses

Category	Analysis
Means and location	One-sample *t* test
	Paired *t* test
	Two-sample *t* test
Confidence intervals	One-sample means
	Paired means
	Two-sample means
Equivalence tests	One-sample means
	Paired means
	Two-sample means
Proportions	One proportion
	Two correlated proportions
	Two independent proportions
Correlation and regression	Pearson correlation coefficient
	Multiple regression
Analysis of variance	One-way ANOVA
Linear models	General linear univariate models
Survival analysis	Two-sample survival rank tests

Features

PSS provides multiple input parameter options, stores the results in a project format, displays power curves, and produces narratives for the results. Narratives are descriptions of the input parameters and a statement about the computed power or sample size. The SAS log and SAS code are also available.

All analyses offer computation of power or sample size. Some analyses offer computation of sample size per group as well as total sample size.

Where appropriate, several alternate ways of entering values for certain parameters are offered. For example, in the two-sample *t* test analysis, means can be entered for individual groups or as a difference. The null mean difference can be specified as a default of zero or can be explicitly entered.

Information about existing analyses is stored in a project format. You can access each project to review the results or to edit your input parameters and produce another analysis.

Chapter 2
A Quick Start

Chapter Contents

Chapter 2
A Quick Start

Overview

This chapter is intended to get you off to a quick start with PSS. More detailed information about using the application is found in Chapter 3, "How to Use PSS," and in the example chapters.

To start the application, invoke the Microsoft Internet Explorer browser. Enter the appropriate Web address for the application. For information on accessing the application and about Web applications, see Appendix A, "Web Applications."

Important Tip

Do not use the **Back** and **Forward** buttons on your Web browser to move from page to page within the application. Doing so bypasses the application and produces unpredictable results. Instead, use the navigation bar, breadcrumb trail, or the buttons on each Web page, as described in Chapter 3, "How to Use PSS."

A Simple Example

Suppose you want to determine the power for a new marketing study. You want to compare car sales in the southeastern region to the national average of 1.0 car per salesperson per day. You believe that the actual average for the region is 1.6 cars per salesperson per day. You want to test if the mean for a single group is larger than a specific value, so the one-sample t test is the appropriate analysis. The conjectured mean is 1.6 and the null mean is 1.0. You intend to use a significance level of 0.05 for the one-tailed test. You want to calculate power for two standard deviations, 0.5 and 0.75, and two sample sizes, 10 and 20 dealerships.

First, you choose the appropriate analysis on the Analyses page. When you invoke the application, either the Projects page or the Analyses page is displayed. If you have used the application before, the Projects page is displayed; otherwise, the Analyses page is displayed.

Figure 2.1.   Navigation Bar with Analyses Page Highlighted

To get to the Analyses page, select **Analyses** in the navigation bar at the top of the page as shown in Figure 2.1.

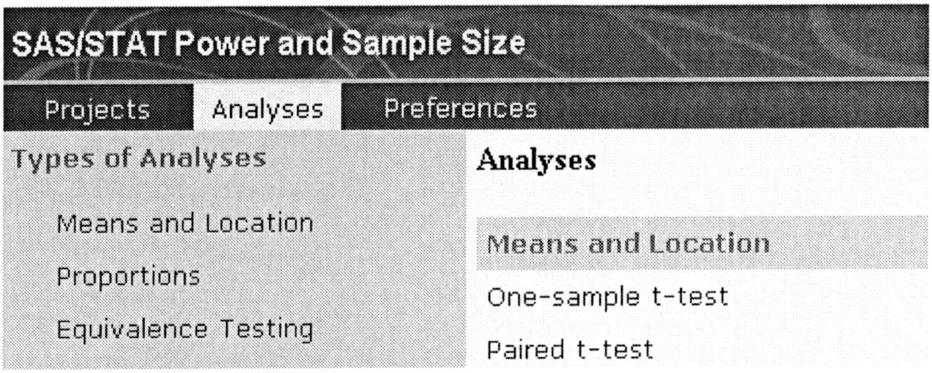

SAS/STAT Power and Sample Size

| Projects | Analyses | Preferences |

Types of Analyses

   Means and Location

   Proportions

   Equivalence Testing

Analyses

   Means and Location

   One-sample t-test

   Paired t-test

Figure 2.2.   Analyses Page with One-sample *t* Test Choice

For this example, the analysis is the `One-sample t-test` in the **Means and Location** section, which is displayed in the lower right portion of Figure 2.2. Select the analysis from the list. When you do, its Input page is displayed. If you want to view help that describes this analysis, select the **Help** link in the upper right portion of the Input page.

Entering Parameter Values

You must choose whether to calculate power or sample size. You can also provide a more descriptive label for this analysis. The description is displayed on the Projects page, enabling you to differentiate the analysis from other ones.

Description: | Regional car sales versus the national average |

| Calculate | Power ▼ | | Save and Close | Reset |

Figure 2.3. Power Selected from Drop-down List

For this example, select **Power** from the drop-down list to the right of the **Calculate** button. Change the description to **Regional car sales versus the national average**. These changes are shown in Figure 2.3.

Now you must provide values for two analysis options and four parameters. These sections of the Input page are labeled Hypothesis, Distribution, Alpha, Mean, Standard Deviation, and Sample Size.

Hypothesis

Because you are only interested in whether the southeastern region produces higher daily car sales the national average, select **One-tailed test** in the **Hypothesis** section, as shown in Figure 2.4.

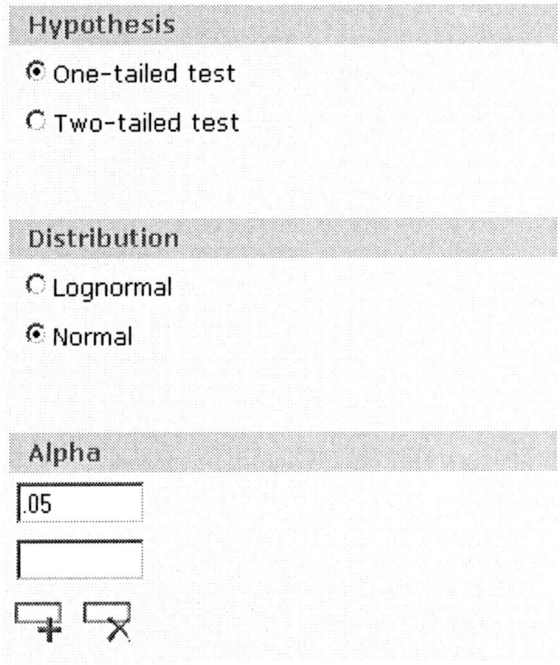

Hypothesis

◉ One-tailed test

○ Two-tailed test

Distribution

○ Lognormal

◉ Normal

Alpha

.05

Figure 2.4.   Hypothesis, Distribution, and Alpha Options

Distribution

You are using means rather than mean ratios, so choose **Normal** in the **Distribution** section, as shown in Figure 2.4.

Alpha

In the **Alpha** section, enter **0.05**, as shown in Figure 2.4. This value will be the default unless it has been changed on the Preferences page.

If you want help for this or other parameters, including the range of valid values, select the context help (**?**) icon on the right side of the corresponding section header, as displayed in Figure 2.5.

Alpha   **?**

Figure 2.5.   Alpha Section Help Icon

Mean

Means can be entered in several different formats, as indicated by the **Select Alternate Forms** link. The list of forms is displayed when you select the link.

For the example, select the **Select Alternate Forms** link in the **Mean** section and then select the `Mean, Null mean` choice from the list. Then, enter `1.6` in the Means table and `1.0` in the Null Means table. Figure 2.6 displays the mean and null mean form with the entered values. Note that additional input rows are available if you want to enter additional sets of parameters.

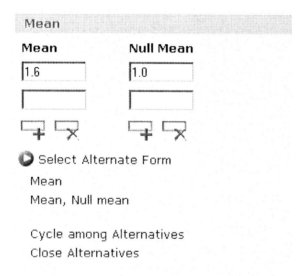

Figure 2.6. Mean and Null Mean Tables with Alternate Forms List Displayed

Standard Deviation

You are interested in two standard deviations, 0.5 and 0.75. Enter them in the table in the **Standard Deviation** section, as shown in Figure 2.7.

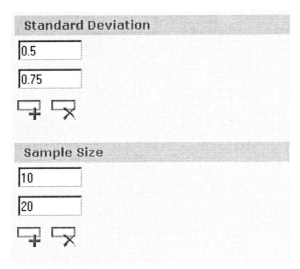

Figure 2.7.   Standard Deviation and Sample Size Tables

Sample Size

You want to be able to sample between 10 and 20 dealerships. Enter these two values in the **Sample Size** table, as shown in Figure 2.7.

Scenarios

Each of the two standard deviations is combined with each of the two sample sizes for a total of four scenarios. Then power is computed for each scenario. In this example, only a single value or setting is present for the mean, null mean, and alpha level, so they are common to all scenarios.

Selecting Results Options

Several results options are available: a Summary Table, Graphs, and Narratives.

For this example, select all three results options: `Summary table of parameters`, `Power by Sample Size` graph, and `Narratives for selected results`, as shown in Figure 2.8.

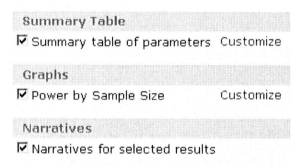

Figure 2.8. Selected Results Options

Customizing the Power by Sample Size Graph

The summary table is created using the two sample sizes specified in the Sample Size table, 10 and 20. If you want to create a graph that contains more than these two sample sizes, you can do so by customizing the graph. Select the **Customize** link beside the Power by Sample Size graph choice item in the **Graphs** section. The Customize Graph dialog page that is subsequently displayed is shown in Figure 2.9.

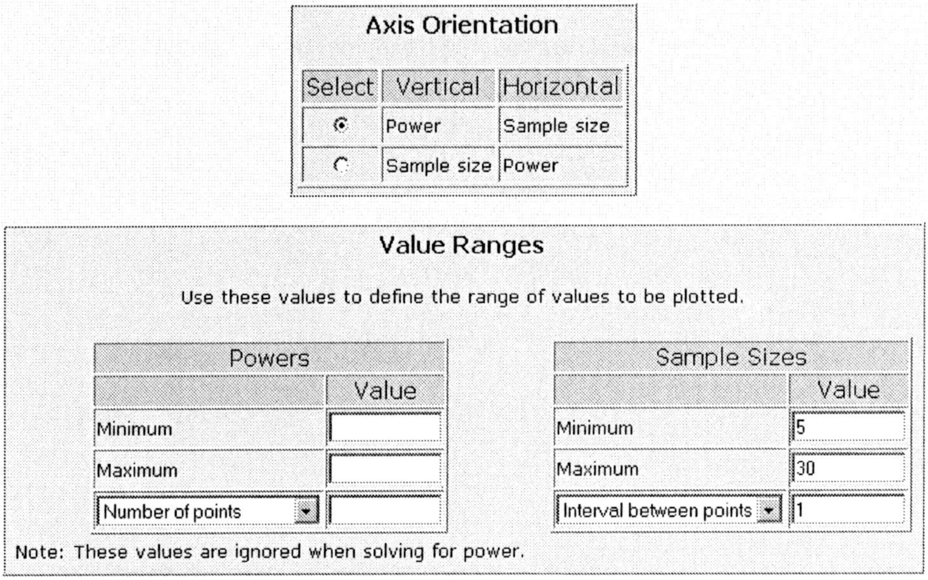

Axis Orientation

Select	Vertical	Horizontal
⦿	Power	Sample size
○	Sample size	Power

Value Ranges

Use these values to define the range of values to be plotted.

Powers	Value		Sample Sizes	Value
Minimum			Minimum	5
Maximum			Maximum	30
Number of points ▾			Interval between points ▾	1

Note: These values are ignored when solving for power.

Figure 2.9. Customize Graph Dialog

In the **Axis Orientation** section you can choose to display power on the vertical or horizontal axis with sample size appearing on the other axis. The default is `Power` on the vertical axis, which is appropriate for this graph.

For the **Value Ranges** section, set the axis range for sample sizes in the Sample Sizes table by entering 5 for the minimum and 30 for the maximum. Also, select `Interval between points` in the drop-down list and enter a value of 1. These values set the sample size axis to range from 5 to 30 in increments of 1. The completed Value Ranges section of the dialog is shown in Figure 2.9.

Note that values entered in the Power table are ignored since you are solving for power. You cannot set the range of axis values for the quantity that you are solving for.

Press the **OK** button to save the values that you have entered and return to the previous page.

Performing the Analysis

You have now specified all of the necessary input values. You can perform the analysis by pressing the **Calculate** button. Alternatively, you could choose to save the information that you have entered by pressing the **Save and Close** button and perform the analysis at another time. No error checking is done when you choose the Save and Close action. You can access an analysis that was closed in this way from the Projects page.

For this example, press the **Calculate** button. A message appears on the page indicating that the page has been submitted, as shown in Figure 2.10. Do not press the **Calculate** button more than once. Otherwise, multiple SAS jobs will be submitted that may interfere with each other.

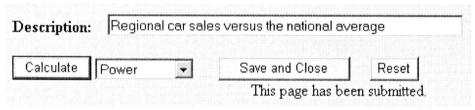

Figure 2.10. "Page Submitted" Message

Viewing the Results

The results that you selected are displayed on the Results page. Use the links in the table of contents on the left side of the page to view the various results.

Summary Table

The Summary table of parameters is composed of two tables, as shown in Figure 2.11. The **Fixed Scenario Elements** table includes the parameters or options that have a single value for the analysis. The **Computed Power** table contains the input parameters that have been given more than one value as well as the computed quantity, power. Thus, the Computed table contains four rows for the four combinations of standard deviation and sample size. From the table you can see that all four powers are high. The smallest value of power, 0.754, is associated with the largest standard deviation and the smallest sample size. In other words, the probability of rejecting the null hypothesis is greater than 75% in all four scenarios.

One-sample t Test for Mean

Fixed Scenario Elements	
Distribution	Normal
Method	Exact
Number of Sides	1
Null Mean	1
Alpha	0.05
Mean	1.6

Computed Power			
Index	Std Dev	N Total	Power
1	0.50	10	0.967
2	0.50	20	>.999
3	0.75	10	0.754
4	0.75	20	0.964

Figure 2.11.   Fixed Scenario Elements and Computed Power Tables

Power by Sample Size Graph

The power by sample size graph in Figure 2.12 contains one curve for each standard deviation. For a standard deviation of 0.5 (the upper curve), increasing sample size above 10 does not lead to much larger power. If you are satisfied with a power of 0.75 or greater, 10 samples would be adequate for standard deviations between 0.5 and 0.75.

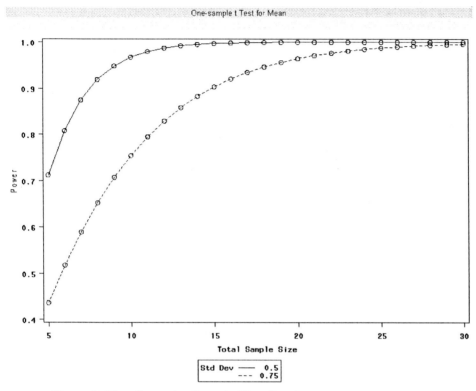

Figure 2.12. Power by Sample Size Graph

Narratives

Narratives are descriptions of the values that compose each scenario and a statement about the computed power or sample size. To view the narratives, select the **Narratives** link in the table of contents on the left of the page. When the narratives result is selected on the Input page, the narrative for the first scenario is created by default:

> For a one-sample *t* test of a normal mean with a one-sided significance level of 0.05 and null mean 1, assuming a standard deviation of 0.5, a sample size of 10 has a power of 0.967 to detect a mean of 1.6.

The narrative that you see is for a standard deviation of 0.5 and a sample size of 10. Suppose that you want to create a narrative for another scenario as well. Select the **Create Narratives** link under **Actions** in the table of contents. A selector table is displayed, as shown in Figure 2.13. Select the third row (the power is 0.754) and leave the first row selected. Press the **Create Narratives** button below the selector table. The Results page is redisplayed. Select the **Narratives** link in the table of contents again.

Obs	Select	Index	Sides	Null Mean	Alpha	Mean	Std Dev	N Total	Power
1	☑	1	1	1	0.05	1.6	0.50	10	0.967
2	☐	2	1	1	0.05	1.6	0.50	20	1.000
3	☑	3	1	1	0.05	1.6	0.75	10	0.754
4	☐	4	1	1	0.05	1.6	0.75	20	0.964

Figure 2.13. Narrative Selector Table

Two narratives are displayed. Note that the two narratives differ for the values of the standard deviation and the power:

For a one-sample *t* test of a normal mean with a one-sided significance level of 0.05 and null mean 1, assuming a standard deviation of 0.5, a sample size of 10 has a power of 0.967 to detect a mean of 1.6.

For a one-sample *t* test of a normal mean with a one-sided significance level of 0.05 and null mean 1, assuming a standard deviation of 0.75, a sample size of 10 has a power of 0.754 to detect a mean of 1.6.

Other Results

Other results include the Input parameters, the SAS log, and the SAS code. The Input parameters are a collection of tables that list the input parameters; the calculated quantity (power, in this example) is not included.

The SAS log that was produced when the Calculate button was last pressed is available from the SAS log link.

The SAS code that produced the results is available from the SAS code link.

Printing Results

Use your browser's print facility to print the results that you want.

Three of the results — the table, the graph, and the narratives — are displayed in the browser on a single page. If printed, all three would appear together. If you want to print one of these results individually, select the Print-friendly version icon (Figure 2.14) beneath the result. The result is displayed in a separate browser window so that it can be printed individually.

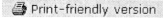

Figure 2.14.   Print-friendly Version Icon

Accessing the Analysis from the Projects Page

You can access previously performed analyses from the Projects page. Select the **Projects** item on the navigation bar. As displayed in Figure 2.15, the analysis that you just completed is listed in the table. The label that you assigned to it, `Regional car sales versus the national average`, appears in the **Description** column of the table. The table also contains the date that the analysis was last modified. If you do not see the analysis, change the **Display analyses by date** selection to `All`, by selecting `All` from the drop-down list and pressing the **Display** button. The analyses are displayed in reverse chronological order.

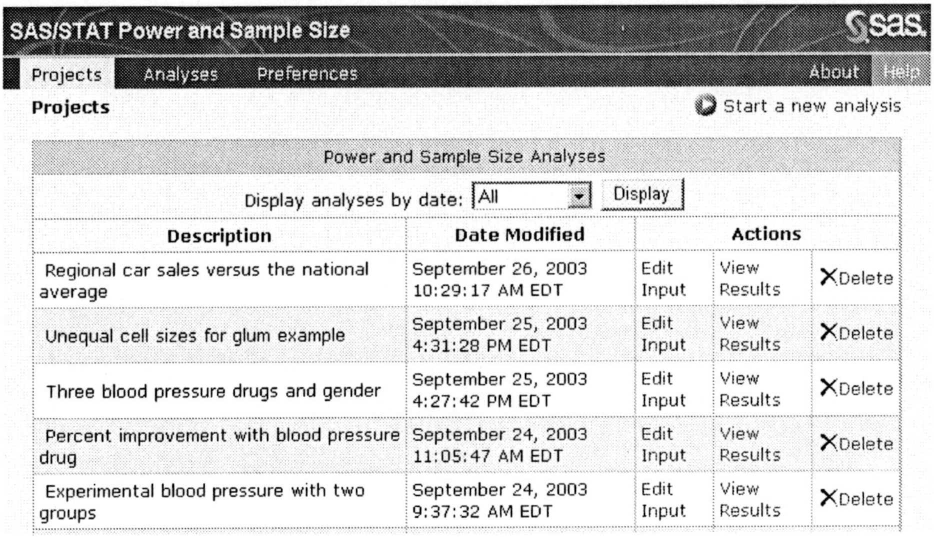

Figure 2.15. The Projects Page Containing the Analysis Created in the Example

Three actions are available with each analysis:

Edit Input

> Open the Input page for this analysis. The previously specified parameters and options are displayed. Use this action to change values or settings and rerun the analysis.

View Results

> Open the Results page for this analysis. The previously generated results are displayed. Use this action to review previous results.

Delete

> Delete all information about the analysis.

For this example, select the **Edit Input** link. The Input page for the analysis is displayed. You can see that all of the information that you previously specified is present.

Changing Values and Rerunning the Analysis

After viewing the graph, you may want to re-create the graph with a different range for sample sizes. In the **Graph** section on the Input page, select the **Customize** link for the power by sample size graph. The Customize Graph dialog page is displayed.

In the **Value Ranges** section of the dialog page, change the **Maximum** value in the Sample Size table from 30 to 20. Press the **OK** button.

Back on the Input page, rerun the analysis by pressing the **Calculate** button. The Results page is redisplayed and the graph now has the new maximum value for the sample size axis.

Chapter 3
How to Use PSS

Chapter Contents

Chapter 3
How to Use PSS

Overview

The PSS application is accessed using the Microsoft Internet Explorer Web browser. Browser-based applications differ from client applications by being organized into Web pages instead of windows and by the way you navigate them.

More information specific to Web applications, including how to access the application, is found in Appendix A, "Web Applications."

Navigating

Within PSS there are several ways to navigate pages and several ways to scroll within a page. The navigation bar and breadcrumb trail enable you to move from page to page. The links in the table of contents and the scroll bar enable you to move within a page.

Important Tip

Do not use the **Back** and **Forward** buttons on your Web browser to move from page to page within the application. Doing so bypasses the application and produces unpredictable results.

Navigation Bar

The *navigation bar* is the principal way of moving from one page to another. It appears in the banner area at the top of the page beneath the application name. As illustrated in Figure 3.1, three pages are available from the navigation bar: the Projects page, the Analyses page, and the Preferences page. The current page is indicated by color shading, as with **Analyses** in Figure 3.1.

Figure 3.1. Navigation Bar

Breadcrumb Trail

Sometimes a breadcrumb trail is present in the banner area beside the application name. A *breadcrumb trail* is a list of navigation links of the pages that you have just visited. The term is derived from using the links to retrace one's steps. It takes the form

<u>**Previous page 2**</u> > <u>**Previous page**</u> > **Current page**

Current page refers to the page that you are on. It is not selectable. **Previous page** refers to the page that was displayed immediately before you arrived at the current page. By selecting it, you return to the previous page. There may be a sequence of previous pages; for example, **Previous page 2**. Figure 3.2 shows a breadcrumb trail from the application.

Figure 3.2. Breadcrumb Trail

Breadcrumb navigation links are displayed when you select **Edit Input** or **View Results** for an existing analysis on the Projects page and when you press the **Calculate** button on the Input page. The use of these pages is described later in this chapter.

Breadcrumb navigation trails are used in PSS to provide links for pages that you might like to navigate to as well as pages that you have actually visited. For example, when you select **View Results** for an existing analysis on the Projects page, both **Analysis Input** and **Analysis Results** appear in the breadcrumb trail. Although you moved directly from the Projects page to the Results page, PSS provides the link to the Input page as a convenience.

An important use of breadcrumb navigation is to enable you to move from a Results page to the corresponding Input page. Neither of these pages appears on the navigation bar.

Table of Contents

Some pages of the PSS application include a table of contents that appears on the left side. Links in the table of contents enable you to move from section to section on the same page. When you select a link, the contents on the right of the page are scrolled to the appropriate section.

In Figure 3.3, the table of contents is the area on the left entitled **Types of Analyses**. If you select the second link, **Proportions**, the list on the right scrolls to the top of the **Proportions** section on the right.

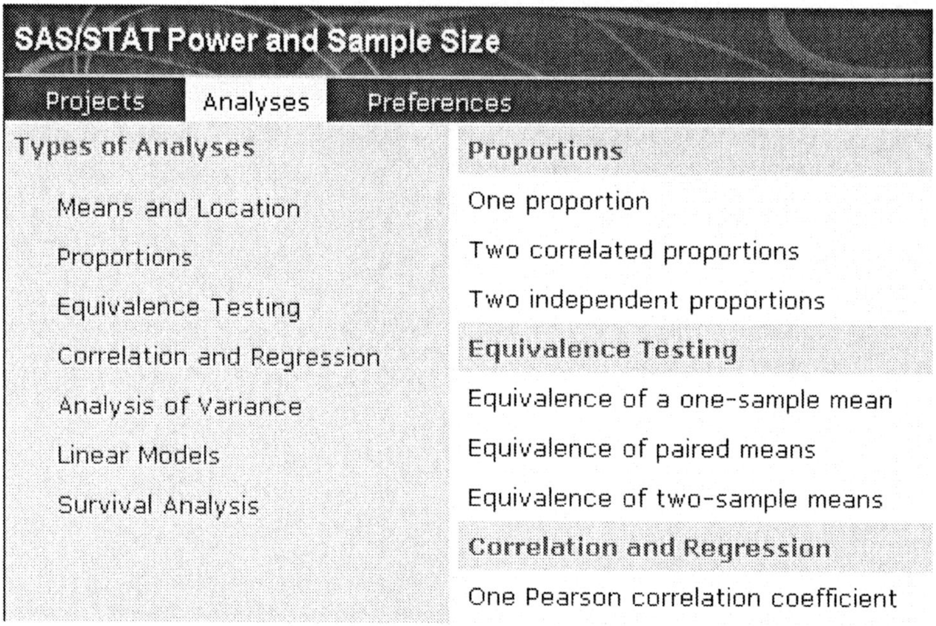

Figure 3.3. Table of Contents on Analyses Page

The scrolling action does not continue below the bottom of the contents portion of the page. If you select a section that is near the bottom of the page, it may not appear at the top of the contents area. Instead, it may appear part of the way down the page.

Pages

As a Web application, the PSS is organized into Web pages. The use of the various pages is described in this section.

Using the Projects Page

The Projects page lists analyses that you previously created. Use this page to manage your previous work. You display the Projects page by selecting **Projects** in the navigation bar at the top of the page. A Projects page is displayed in Figure 3.4.

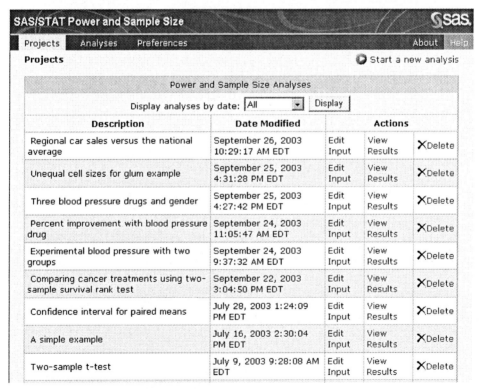

Figure 3.4. Projects Page

The table of existing analyses contains the **Description** of the analysis, its last **Date Modified**, and several possible **Actions**: **Edit Input**, **View Results**, and **Delete**. The analyses are displayed in reverse order of the dates that they were last modified. You set the label that appears in the **Description** column on the Input page discussed below. Choose a description that differentiates each analysis from others.

You can choose to display a subset of the analyses based on one of several ranges of the date that the analysis was last modified. Select the desired date range in the **Display analyses by date** drop-down list and press the **Display** button. Some of the available date ranges are `All`, `Today`, `Yesterday`, `This week`, and `This year`.

To change the input parameters of an existing analysis, select the **Edit Input** action for the desired analysis. The Input page for this analysis is displayed.

To view the results of an existing analysis, select the **View Results** action for the desired analysis. The Results page for this analysis is displayed. A breadcrumb navigation trail to the Analysis Input for this analysis is also displayed.

To delete an existing analysis, select the **Delete action**. You are prompted for a confirmation of the deletion. All information about the analysis is deleted.

In addition to using the navigation bar to display the Analyses page, you can also select **Start a new analysis** at the upper right corner of the page.

Using the Analyses Page

The Analyses page lists the available analyses and enables you to open a new analysis. You display the Analyses page by selecting **Analyses** in the navigation bar at the top of the page. Figure 3.5 shows part of the Analyses page.

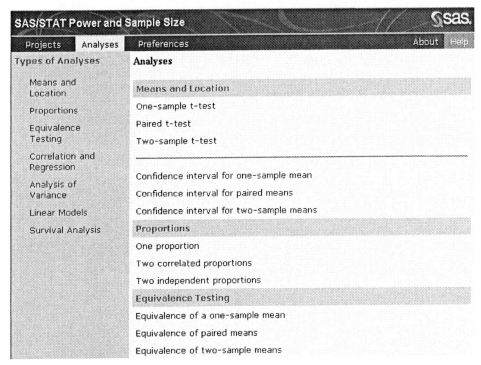

Figure 3.5. Analyses Page

To create a new instance of an analysis, select its name in the list. For example, to perform an analysis for a one-sample *t* test, select `One-sample t test` from the list. When you select an analysis, the Input page for that analysis is displayed.

Use the table of contents on the left side of the page to scroll among the analyses.

Using the Preferences Page

On the Preferences page you can set default values for options that are used by all analyses. You display the Preferences page by selecting **Preferences** in the navigation bar at the top of the page. Figure 3.6 shows the Preferences page.

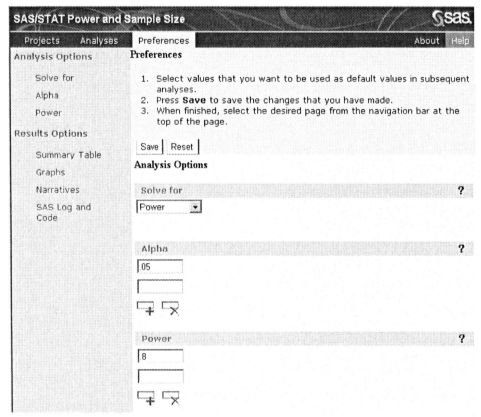

Figure 3.6. Preferences Page

These preference values are used as the defaults for each newly opened analysis, that is, those that are opened from Analysis page. Changes in preferences on this page do not change the state of an existing analysis, that is, one that is accessed from the Projects page.

After changing preference values, press the **Save** button to save the changes. You can then leave the Preferences page. If you want to reset all values to their state when they were last saved, press the **Reset** button.

Setting Analysis Options

You can set defaults for the quantity to be solved for (power or sample size), alpha (the significance level), and power. Each of these default values can be overridden on the Input page for a specific analysis.

To set the default quantity to be solved for, select `Power` or `Sample size` from the drop-down list in the **Solve for** section. For analyses that produce more than one type of sample size, the `Sample size` selection on this page corresponds

to total sample size. For analyses like the confidence intervals of means, selecting `Power` is equivalent to selecting `Prob(Width)`.

To set default values of alpha, enter one or more values in the **Alpha** data entry table. It is not necessary to have any default values for alpha. Add more rows to the table as needed using the Add a Row control (**+**) control at the bottom of the table.

To set default values of power, enter one or more values in the **Power** data entry table. It is not necessary to have any default values for power.

Setting Results Options

You can also set default selections for the Summary Table, the Graphs, Narratives, and the SAS Log and Code choices.

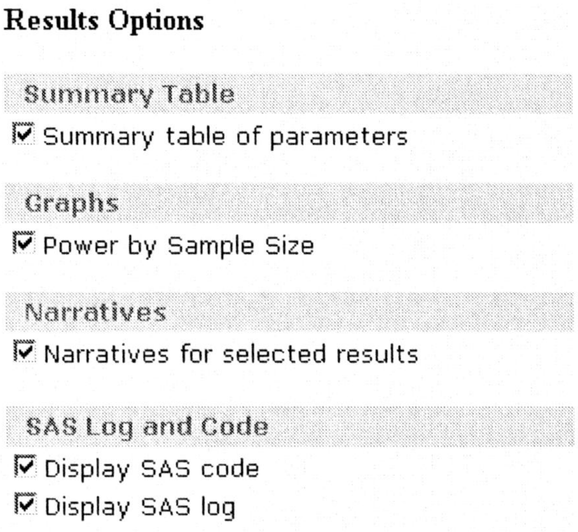

Results Options

Summary Table
☑ Summary table of parameters

Graphs
☑ Power by Sample Size

Narratives
☑ Narratives for selected results

SAS Log and Code
☑ Display SAS code
☑ Display SAS log

Figure 3.7. Results Options Preferences

The **Summary table of parameters** consists of the input parameter values and the calculated quantity (power or sample size). Select the corresponding check box to enable the creation of this result as the default.

To request that an analysis creates a **Power by Sample Size** graph by default, select the corresponding check box.

The **Narratives for selected results** are descriptions of the input parameter values and calculated quantities in sentence or paragraph form. Each narrative corresponds to one calculated quantity value. Select the corresponding check box to enable the creation of this result as the default.

The SAS code used to generate the analysis results and the resulting SAS log are always created and saved. Use the **Display SAS code** and **Display SAS log** check boxes to control whether these results can be viewed from the Results page.

Help for the page is displayed by selecting the **Help** link at the top right of the page.

The context help icon (**?**) on the right side of each section header (the shaded bar) provides access to help for the corresponding option.

Using the Input Page

When you open a new analysis on the Analyses page or select the **Edit Input** link for an existing analysis on the Projects page, the Input page for that analysis is displayed. It is not directly accessible from the navigation bar. Use this page to specify the values of all input parameters as well as to select any desired results. Figure 3.8 shows an instance of an Input page.

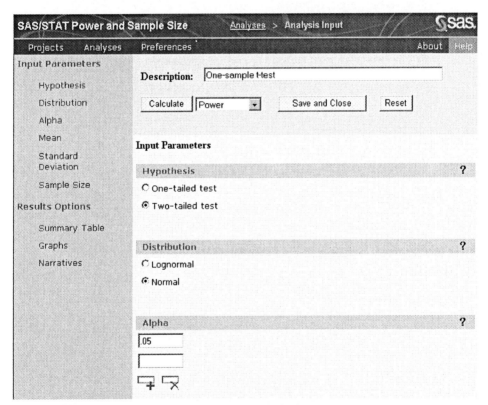

Figure 3.8. The Input Page for an Analysis

Description

The description is displayed on the Projects page to enable you to identify this particular analysis. By default it is the name of the analysis.

Quantity to Solve For

Select the quantity to solve for, power or sample size, from the drop-down list that is located below the description and to the right of the **Calculate** button.

Actions

Several actions are available, which correspond to the **Calculate**, **Save and Close**, and **Reset** buttons.

At any point when you are entering information, you can choose to save the information by pressing the **Save and Close** button. The information is saved but data validation is not performed. This action is useful when you have not finished entering information and want to continue at another time. The Input page is closed and the Projects page is displayed.

After providing all necessary information, you can choose to perform the analysis by pressing the **Calculate** button. Data validation is performed. If errors are found, the Input page is redisplayed. If no errors are found, the Results page is displayed. When you press the **Calculate** button, a message is displayed saying that the page has been submitted. Do not press the button more than once.

The **Reset** button enables you to reset parameters and options to their previous value when the Input page was last displayed. If you choose to display a dialog page (see the "Using Dialog Pages" section on page 41), the values and settings on the Input page are saved. If you press the Reset button on the Input page after returning from the dialog page, the fields and selections are set to the last values saved — those in effect when you left the Input page to open the dialog page.

Discarding Changes

There are two ways to discard changes made on the Input page. You can press the **Reset** button, which restores values to their state when the page was last displayed (as discussed above). You can also move to another page, such as the Analyses or Projects page. Changes made to the Input page are not saved.

Data Entry and Options Controls

The Input page contains various controls by which you can enter values or select choices. In addition to the usual data entry controls such as text fields and check boxes, several specialized controls are present — data entry tables and the Select Alternate Form facility. More detailed descriptions follow.

Data Entry Tables

Data entry tables are composed of data entry fields for one or more rows and columns. Figure 3.9 shows a two-row, two-column table.

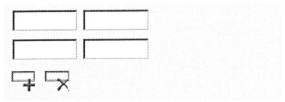

Figure 3.9. Two-Column Data Entry Table with Controls

Enter the appropriate value in each field. It is not necessary to enter data in all rows or to delete empty rows. However, if a table has more than one column, the fields of a row must be completely filled or completely blank. Rows with values in some fields but not all fields are not allowed.

To add more rows, press the Add a Row control (+) beneath the table. To delete the last row, press the Delete Last Row control (–).

Selecting Alternate Forms

For some input parameters, there are several ways in which data may be entered. For example, in the Two-sample *t* test analysis, group means can be entered as either individual means or a difference between means.

For those parameters that have alternate data entry forms, the **Select Alternate Form** link is displayed beneath the table, as shown in Figure 3.10.

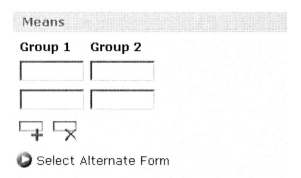

Figure 3.10. Select Alternate Form Control before Opening

Selecting the link displays the list of alternate forms. You can either choose one of the forms or repeatedly press the **Cycle among Alternates** choice to view each form sequentially, as shown in Figure 3.11. Press **Close Alternates** to hide the list. Figure 3.11 shows the second form, **Differences between groups**, selected.

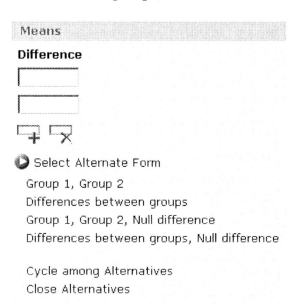

Figure 3.11. Select Alternate Form Control after Opening

For your convenience, the last used alternate form for an analysis is saved and employed as the default when a new instance of the analysis is opened.

Validation and Error Messages

For data entry fields, data validation is performed when you change the focus to another field. For example, if you enter a negative probability for an alpha value, when you move the cursor to another field, the value is checked and an error message is displayed. Additional validation is performed when the **Calculate** button is pressed.

Error messages are displayed in two ways. For some errors a message box appears overlying the browser, as shown in Figure 3.12. The message describes the nature of the error. In some cases, you are asked if you want to proceed with the analysis.

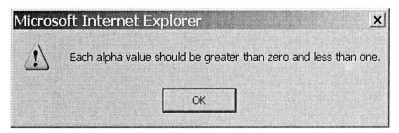

Figure 3.12. Error Message Appearing in Message Box

Other error messages appear as text on the page, as shown in Figure 3.13. The section containing the error is marked both in the table of contents and in the content area. These errors are ones that have been detected when the **Calculate** button is pressed. Correct the problem and press the **Calculate** button again.

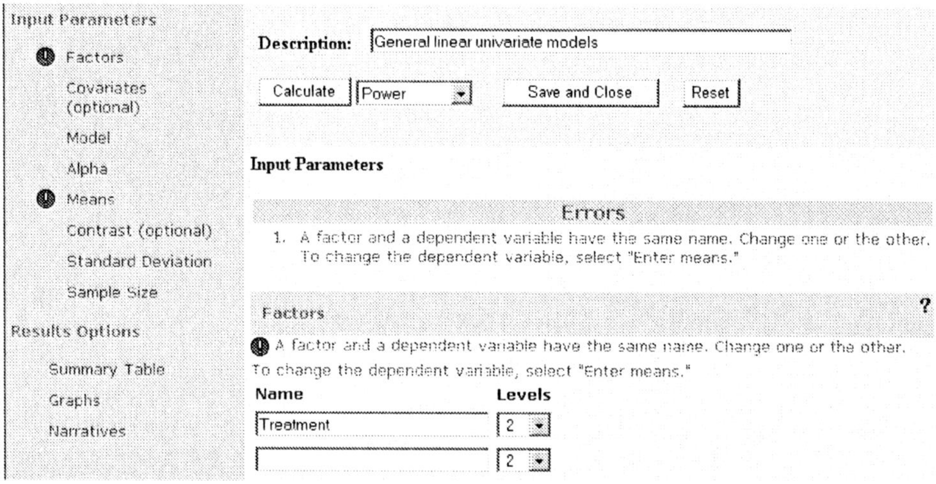

Figure 3.13. Error Message Appearing on a Page

Help

Help for the page is displayed when you press the **Help** link in the upper right corner of the page.

Also, help about each input parameter or results option is available. Each parameter or option section contains a brown shaded header bar that extends across the page. Selecting the context help (**?**) icon at the right side of the bar displays help about this parameter in a separate browser.

Using Dialog Pages

In some cases, data values and selections are displayed on a secondary page called a dialog page. To access the dialog page, select the link on the Input page. For example, as shown in Figure 3.14 from the One-way ANOVA analysis, the **Enter means** link appears in the **Means** section.

Means
Enter means

Figure 3.14. The Enter Means Link to Access Means Dialog Page

To display a data entry form for the means, select the link and the Enter Means dialog page is displayed, as shown in Figure 3.15:

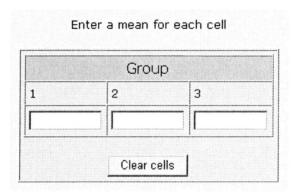

Figure 3.15. One-way ANOVA Means Dialog

All dialog pages have **OK** and **Cancel** buttons and some have other controls as well. After entering data on a dialog page, press the **OK** button to save the data and return to the previous page, or press the **Cancel** button to discard the data and return to the previous page. Pressing the **Reset** button restores values and selections to their state when the page was initially displayed.

Customizing Graphs

One example of a dialog page is the Customize Graph page. This page is available for all analyses. It is described in detail in this section.

The page consists of title fields, an **Axis Orientation** section and a **Value Ranges** section. Use the **Axis Orientation** choices to specify which axes you want used for power and for sample size. Use the **Value Ranges** section to specify the axis range for the nontarget quantity, that is, the power axis if you are solving for sample size or vice versa.

You cannot set the axis range for the quantity being computed. A note to this effect is displayed beneath the corresponding table. For example, when solving for power, a note appears beneath the Power table that those values will be ignored.

When specifying a value range, specify a minimum value and a maximum value. Also, select either the `Number of points` or the `Interval be-`

`tween points` choice for the axis and specify a value. All of these values are optional; use only the ones you want.

Using the Results Page

The Results page for an analysis is displayed when you have pressed the **Calculate** button on the Input page or selected the **View Results** action for an existing analysis on the Projects page. It is not directly accessible from the navigation bar. Use this page to view and work with the results for this analysis. Figure 3.16 shows a Results page for an analysis.

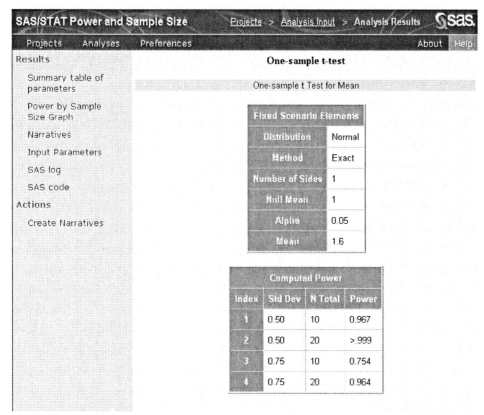

Figure 3.16. Results Page for an Analysis

A breadcrumb navigation trail is displayed in the banner at the top of the page. The right-most item in the breadcrumb trail, entitled **Analysis Results**, corresponds to the current page, and the item to its left, entitled **Analysis Input**, corresponds to the Input page for the analysis. The breadcrumb trail links enables you to quickly return to the Input page to make changes to the analysis. The banner also contains the help link for this page.

The **Summary table of parameters**, **Power by Sample Size** graph, and **Narratives** are displayed in the contents area. You can use the table of contents links or the content area scrollbar to scroll among them.

The **Summary table of parameters** consists of two tables, as shown in Figure 3.16. The **Fixed Scenario Elements** table includes the options and parameter values that are constant for the analysis. The **Computed** table includes the calculated power or sample size values and the values for input parameters that have multiple values specified for the analysis.

Beneath each table, graph, or set of narratives is a **Print-friendly version** link. Selecting this link for one of these three results causes that particular result to be displayed by itself in a separate browser window. Use this feature to print one of the results rather than all three together.

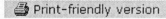

Figure 3.17. Print-friendly Version Icon and Link

Other results include a set of tables containing the values of the input parameters, the SAS log and SAS code (if their display preferences have been selected on the Preferences page), and the **Create Narrative** action that provides access to the Create Narrative facility. Use the appropriate links in the table of contents on the left to display these results.

Creating Narratives

When narratives are requested, only the narrative for the first scenario (the first row of the summary table) is created. Others can be requested from the Results page using the Create Narrative facility. Select the **Create Narratives** link in the **Actions** section of the table of contents (shown in Figure 3.18). Even if you do not request narratives on the Results page, you can generate them using the Create Narrative facility.

Figure 3.18. The Create Narrative Link to Access the Create Narrative Facility

When you select the **Create Narratives** link, a selection table is displayed that contains each combination of input parameters and the corresponding calculated target quantity. Select the check box for each scenario for which you want to generate a narrative. Figure 3.19 shows the selection table with two selected narratives. Then press the **Create Narratives** button. The Results page is redisplayed with the new narratives.

Obs	Select	Index	Sides	Null Mean	Alpha	Mean	Std Dev	N Total	Power	Error	Information
1	☑	1	1	1	0.05	1.6	0.50	10	0.967		
2	☐	2	1	1	0.05	1.6	0.50	20	1.000		
3	☑	3	1	1	0.05	1.6	0.75	10	0.754		
4	☐	4	1	1	0.05	1.6	0.75	20	0.964		

Create Narratives

Figure 3.19. The Create Narrative Selector Table

Chapter 4
Two-Sample t Test

Chapter Contents

Chapter 4
Two-Sample t Test

Overview

The one-sample t test compares the mean of a sample to a given value. The two-sample t test compares the means of two samples. The paired t test compares the mean of the differences in the observations to a given number. PSS provides power and sample size computations for all of these types of t tests. For more information about power and sample size analysis for t tests, see Chapter 57, "The POWER Procedure" (*SAS/STAT User's Guide*).

The two-sample t test tests for differences or ratios between means for two groups. The groups are assumed to be independent. This chapter describes three examples using the two-sample t test: for equal variances, for unequal variances, and for mean ratios.

Test of Two Independent Means for Equal Variances

You are interested in testing whether an experimental drug produces a lower systolic blood pressure than a placebo does. Will 25 subjects per treatment group yield a satisfactory power for this test? From previous work, you expect that the blood pressure is 132 for the control group and 120 for the drug treatment group, and that the standard deviation is 15 for both groups. You want to use a one-tailed test with a significance level of 0.05. Because there are two independent groups and you are assuming that blood pressure is normally distributed, the two-sample t test is an appropriate analysis.

You start by displaying the Analyses page by selecting the **Analyses** item on the navigation bar. Under the **Means and Location** section, select **Two-sample t-test**. The Input page for the analysis is displayed.

Entering Input Parameters

Description

The description is used to identify this particular analysis in the table on the Projects page. It is displayed above the **Calculate** button.

For this example, change the description to `Experimental blood pressure drug with two groups`, as shown in Figure 4.1.

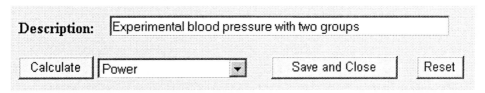

Figure 4.1. Project Description

Solve For

For the two-sample *t* test analysis you can choose to solve for power, total sample size, or sample size per group. Select the desired quantity from the drop-down list to the right of the **Calculate** button.

If the default value for the **Solve For** quantity is not `Power`, change it by selecting `Power` from the drop-down list, as shown in Figure 4.1. The list contains other values that you could solve for, including total sample size and sample size per group. For more information on solving for sample size, see the "Solving for Sample Size" section on page 71.

Hypothesis

You can choose either a one- or two-tailed test. For the one-tailed test, the alternative hypothesis is assumed to be in the same direction as the effect. If you do not know the direction of the effect, that is, whether it is positive or negative, the two-tailed test is appropriate. If you know the effect's direction, the one-tailed test is appropriate. If you specify a one-tailed test and the effect is in the unexpected direction, the results of the analysis are invalid.

Because you are only interested in whether the experimental drug lowers blood pressure, select the `One-tailed test` choice in the **Hypothesis** section, as shown in Figure 4.2.

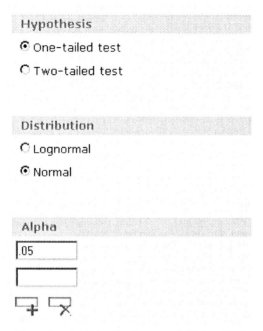

Figure 4.2. Hypothesis, Distribution, and Alpha Options

Distribution

The distribution option specifies the underlying distribution for the test statistic.

For this example, you are interested in means rather than mean ratios, so select the **Normal** choice in the **Distribution** section, as shown in Figure 4.2.

Alpha

Alpha is the significance level that is the probability of falsely rejecting the null hypothesis. If you frequently use the same value(s) for alpha, set them as the default on the Preferences page.

Enter the desired significance level of 0.05 into the first cell of the Alpha table, if it is not already the default value.

Means

For this analysis, you can enter the means for the two groups either individually or as a difference. If your null mean difference is not zero, you can also enter that value.

In the **Means** section, there are four possible ways to enter the group means and the null mean difference. Examine the available alternate forms by selecting the **Select Alternate Form** link. The four available forms are:

Group 1, Group 2

> Enter the means for each group. The null mean difference is assumed to be 0. The difference is formed by subtracting the mean for group 1 from the mean for group 2.

Differences between groups

> Enter the difference between the group means. The null mean difference is assumed to be 0.

Group 1, Group 2, Null difference

> Enter the means for each group and a null mean difference. The difference is formed by subtracting the mean for group 1 from the mean for group 2.

Differences between groups, Null difference

> Enter the difference between the group means and a null mean difference.

For this example, a null mean difference of 0 is reasonable, so select the **Group 1, Group 2** means form from the list, as shown in Figure 4.3. Enter the control mean of 132 in the first row of the first column and the experimental mean of 120 in the first row of the second column.

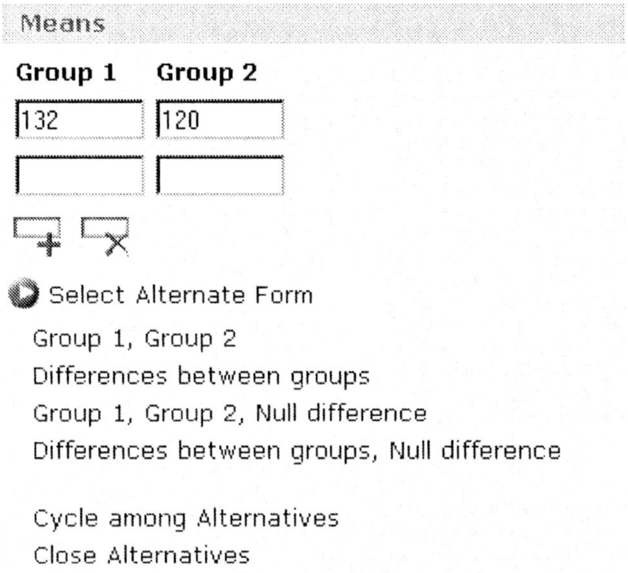

Figure 4.3. Alternative Means Forms and Values

Standard Deviation

In the **Standard Deviation** section, there are two alternate forms for entering standard deviation:

`Common std.   deviation`
>Enter the standard deviation for the two groups. It is assumed to be equal for both groups.

`Group 1, Group 2`
>Enter the standard deviation for each group. The group values may be equal or unequal.

For the example, select the `Common std.   deviations` alternate form in the **Standard Deviation** section and enter a single value of 15, as shown in Figure 4.4.

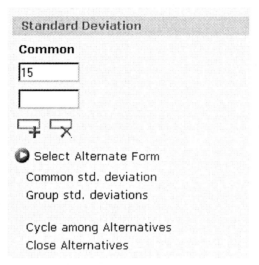

Figure 4.4. Alternative Standard Deviation Forms and Values

Sample Size

In the **Sample Size** section, there are three alternate forms for entering sample sizes:

N per group
> Enter the sample size for the two groups. The group sizes are assumed to be equal.

Group 1, Group 2
> Enter the sample size for each group. The group sizes may be equal or unequal.

Total N, Group weights
> Enter the total sample size for the two groups and the relative sample sizes for each group. For more information on using relative sample sizes, see the "Using Unequal Group Sizes" section on page 71.

Examine the alternatives by selecting the **Select Alternate Form** link. Select the **N per group** form. You want to examine a curve of powers in the

Power by Sample Size graph, so enter the values 20, 25, and 30 in the Sample Size table, as shown in Figure 4.5. If more rows need to be added to the table, add them pressing the Add a Row (+) icon below the table.

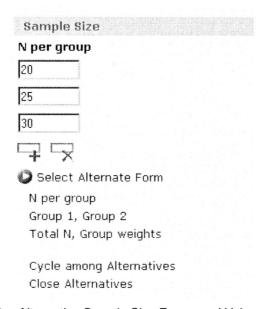

Figure 4.5. Alternative Sample Size Forms and Values

Summary of Input Parameters

Table 4.1 contains the values of the input parameters for the example.

Table 4.1. Summary of Input Parameters

Parameter	Value
Hypothesis	One-tailed test
Distribution	Normal
Alpha	0.05
Means form	Group 1, Group 2
Means	132, 120
Standard deviation form	Common std. deviations
Standard Deviation	15
Sample size form	N per group
Sample Size	20, 25, 30

Results Options

Request all three results by selecting the check boxes for the `Summary table of parameters`, the `Power by Sample Size` graph, and `Narratives for selected results`.

Press the **Calculate** button to perform the analysis. If there are no errors in the input parameter values, the Results page appears. If there are errors in the input parameter values, you will be prompted to correct them.

Viewing Results

The results are listed in the table of contents on the left of the page. Select each result that you want to view.

Summary Table

The Summary table of parameters includes the values of the input parameters and the computed quantity (in this case, power). The table consists of two subtables, the **Fixed Scenario Elements** table that contains the input parameters that have only one value for the analysis, and the **Computed Power** table that contains the input parameters that have more than one value for the analysis and the corresponding power. Only the N per group parameters appears in the Computed table; all of the input parameters have a single value. The computed power for a sample size per group of 25 is 0.874. You have a probability of 0.87 that the study will find the expected result if the assumptions and conjectured values are correct.

Two-sample t Test for Mean Difference

Fixed Scenario Elements	
Distribution	Normal
Method	Exact
Number of Sides	1
Alpha	0.05
Group 1 Mean	132
Group 2 Mean	120
Standard Deviation	15
Null Difference	0

Computed Power		
Index	N Per Group	Power
1	20	0.799
2	25	0.874
3	30	0.922

Figure 4.6. Summary Table

Power by Sample Size Graph

The Power by Sample Size graph displays power on the vertical axis and sample size per group on the horizontal axis. The range of values for the horizontal axis is 20 to 30, which were the largest and smallest values that you entered. In this example, you can customize the graph by specifying the values for the sample size axis (see the "Customizing Graphs" section on page 42).

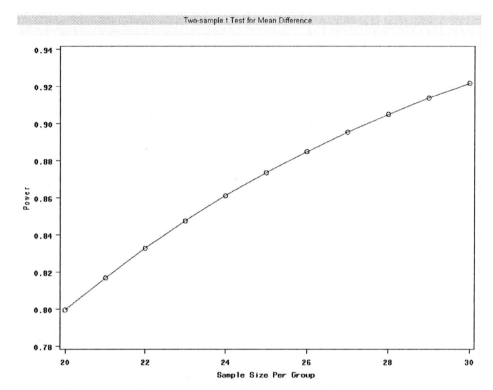

Figure 4.7. Power by Sample Size Graph

Narratives

The Narratives result consists of a sentence- or paragraph-length text summary of the input parameter values and the computed quantity for a single combination of the input parameter values. When narratives are selected as a result option on the Input page, the narrative for the first row in the summary table is generated. In this example, the narrative is for the sample size per group of 20, which yields a power of 0.799:

> For a two-sample pooled *t* test of a normal mean difference with a one-sided significance level of 0.05, assuming a common standard deviation of 15, a sample size of 20 per group has a power of 0.799 to detect a difference between the means 132 and 120.

If you want to create a narrative for another combination of input values, select **Create Narratives** in the **Actions** section of the table of contents. The narrative selector table is displayed. Notice that the first row is selected. To generate other narratives, select the check box that corresponds to the desired row. Here, select the second row for the power corresponding to a sample size of 25. Then press the **Create Narratives** button beneath the selector table. Because you did not deselect the first row, two narratives are produced. The second one for a sample size of 25 lists the power you are interested in, 0.874:

> For a two-sample pooled *t* test of a normal mean difference with a one-sided significance level of 0.05, assuming a common standard deviation of 15, a sample size of 20 per group has a power of 0.799 to detect a difference between the means 132 and 120.

> For a two-sample pooled *t* test of a normal mean difference with a one-sided significance level of 0.05, assuming a common standard deviation of 15, a sample size of 25 per group has a power of 0.874 to detect a difference between the means 132 and 120.

You have finished viewing the results for this example. To change some values of the analysis and rerun it, select the **Analysis Input** link in the breadcrumb navigation items at the top of the page.

Test of Two Independent Means for Unequal Variances

In the example above, you assumed that the population standard deviations were equal. If you believe that the population standard deviations are not equal, use the same two-sample *t* test analysis as above, except change the way that you enter the standard deviations.

You can use the previous example to demonstrate this test. Go to the Projects page by selecting **Projects** on the navigation bar. Identify the analysis that you just performed in the list and select the **Edit Input** link.

Entering Input Parameters

Specifying Group Standard Deviations

Scroll down to the **Standard Deviation** section by using the scroll bar on the right-hand side of the page or clicking on the `Standard Deviation` link in the table of contents on the left. Open the list of alternate forms by selecting the **Select Alternate Form** link. Select the `Group std. deviations` form from the list. Enter the group standard deviations of 12 and 15, as shown in Figure 4.8.

With group standard deviations, the method for the Satterthwaite *t* test is used.

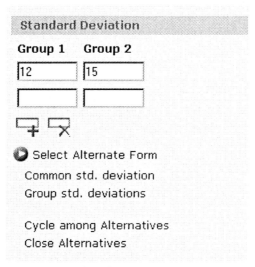

Figure 4.8. Group Standard Deviations

Summary of Input Parameters

Table 4.2 contains the values of the input parameters for the example.

Table 4.2. Summary of Input Parameters

Parameter	Value
Hypothesis	One-tailed test
Distribution	Normal
Alpha	0.05
Means form	Group 1, Group 2
Means	132, 120
Standard Deviation form	Group 1, Group 2
Standard Deviation	12, 15
Sample size form	N per group
Sample Size	20, 25, 30

Rerun the analysis by pressing the **Calculate** button.

Viewing Results

If you modified the previous example, a message window is displayed saying that "Previously selected narratives have been cleared because some input parameter values have been changed." In the previous analysis, you selected two scenarios for which narratives were created. Because you changed from Common to Group standard deviations, those selected narratives were cleared. Use the narrative selector through the Create Narratives action to generate other narratives.

The power for a sample size per group of 25 is 0.924. Notice that the actual alpha is 0.0499. This is due to the Satterthwaite t test being (slightly) biased.

Test of Mean Ratios

Instead of comparing means for a control and drug treatment group, you may want to investigate whether the blood pressure of the treatment group is lowered by a given percentage of the control group, say 10 percent. That is, you expect the ratio of the treatment group to the control group to be 90% or less.

PSS provides a two-sample test of a mean ratio when the data are lognormally distributed.

For mean ratios the coefficient of variation, CV, is used instead of standard deviation. You expect the CV to be between 0.5 and 0.6. You also want to compare an equally weighted sampling of groups with an overweighted one in which the control group contains twice as many subjects in the control group as the treatment group — 50 and 25, respectively.

Entering Input Parameters

Several of the input parameters for the test of mean ratios are the same as the ones described in the the "Test of Two Independent Means for Equal Variances" section on page 49. Mean ratios and coefficients of variation are used instead of mean differences and standard deviations. These two parameters are discussed in detail in this section. For the input parameters and options that have been discussed previously in this chapter, only the values for this example are given.

Enter **Percent improvement with blood pressure drug** as the **Description**, as shown in Figure 4.9.

Select **Power** as the quantity to be solved for from the drop-down list to the right of the **Calculate** button, as shown in Figure 4.9.

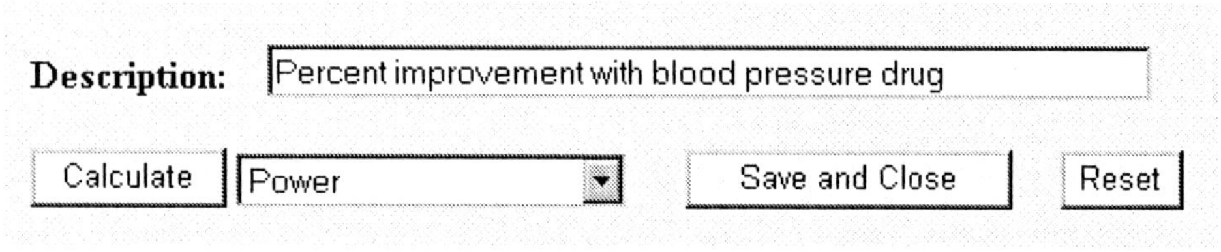

Description: | Percent improvement with blood pressure drug |

Calculate | Power ▼ | | Save and Close | | Reset |

Figure 4.9. Description, Solve for Power

Select the **one-tailed test** choice in the **Hypothesis** section, as shown in Figure 4.10.

You are interested in mean ratios rather than means, so select the **Lognormal** choice in the **Distribution** section, as shown in Figure 4.10.

Enter 0.05 as the significance level into the first cell of the **Alpha** table.

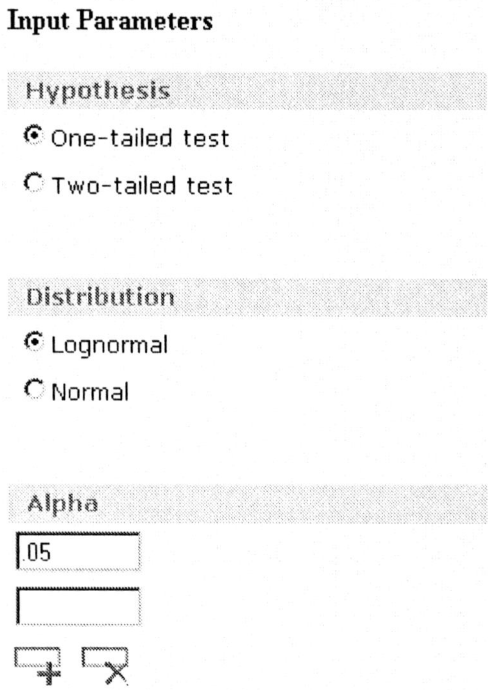

Input Parameters

Hypothesis

⦿ One-tailed test

○ Two-tailed test

Distribution

⦿ Lognormal

○ Normal

Alpha

| .05 |

| |

Figure 4.10. Hypothesis, Distribution, Alpha Options

In the **Sample Size** section, select the second form, `Group 1, Group 2` to enter the sample sizes for each group, and enter values of 25 and 25 in the first row and 25 and 50 in the second row, as shown in Figure 4.12.

Means

In the **Means** section, there are four alternate forms for entering means or mean ratios:

`Group 1, Group 2`
>
> Enter the means for each group. The ratio of the means is formed by dividing the mean for group 2 by the mean for group 1. The null ratio is assumed to be 1.

`Mean ratio`
>
> Enter the ratio of the two group means, that is, the treatment mean divided by the reference mean. The null ratio is assumed to be 1.

`Group 1, Group 2, Null ratio`
>
> Enter the means for each group. The ratio of the means is formed by dividing the mean for group 2 by the mean for group 1. Enter the null ratio.

`Mean ratio, Null ratio`
>
> Enter the ratio of the two group means, that is, the treatment mean divided by the reference mean. Enter the null ratio.

As shown in Figure 4.11, select the second form, `Mean ratio`, which uses a default null ratio of 1. Enter a single value of 0.9.

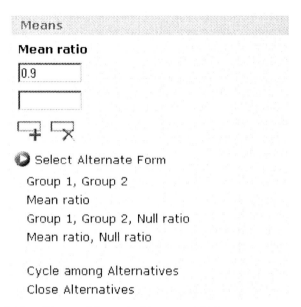

Figure 4.11. Means Alternate Form and Values

Coefficient of Variation

In the **Coefficients of Variation** section, enter the coefficient of variation. It is assumed to be equal for the two groups.

For this example, enter 0.5 and 0.6, as shown in Figure 4.12.

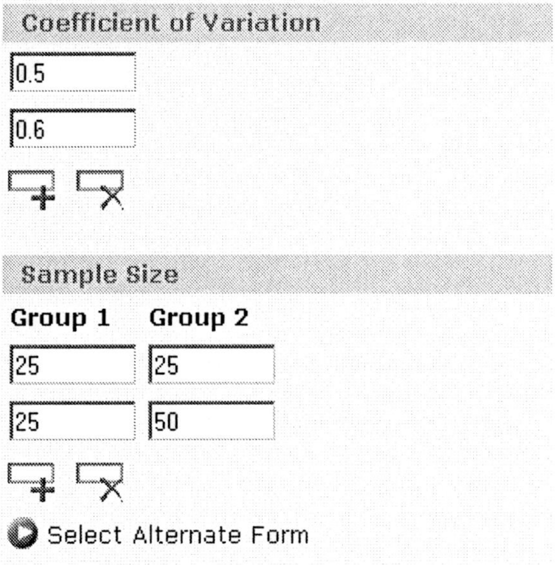

Figure 4.12. Coefficient of Variation and Sample Size

Summary of Input Parameters

Table 4.3 contains the values of the input parameters for the example.

Table 4.3. Summary of Input Parameters

Parameter	Value
Hypothesis	One-tailed test
Distribution	Lognormal
Alpha	0.05
Means form	Mean ratio
Mean ratio	0.9
Coefficients of Variation form	Common cvs.
Coefficients of Variation	0.5, 0.6
Sample size form	Group 1, Group 2
Sample Size	25 25, 25 50

Results Options

Request the `Summary table of parameters` and the `Power by Sample Size` graph in the **Results Options** section.

Press the **Calculate** button to perform the analysis.

In this case, a message dialog is displayed. It informs you that "The power by sample size group is not available when specifying sample sizes for two groups." If you want a power by sample size plot, you can choose to plot total sample size instead by using the `Total N, Group weights` alternate form for sample size. However, you would need to run the analysis separately for the two sets of group sample sizes, since only one set of weights is allowed at a time. For more information on using this input form, see the "Using Unequal Group Sizes" section on page 71.

Viewing Results

The first thing that you notice from the Summary Table in Figure 4.13 is that the calculated powers are quite low — they range from 0.16 to 0.229. You have less than a 25% probability of detecting the difference that you are looking for. Clearly, this set of parameter values leads to insufficient power. To increase power, you might choose a larger sample size or a larger alpha.

Two-sample t Test for Mean Ratio	
Fixed Scenario Elements	
Distribution	Lognormal
Method	Exact
Number of Sides	1
Alpha	0.05
Geometric Mean Ratio	0.9
Null Geometric Mean Ratio	1

Computed Power				
Index	**CV**	**N1**	**N2**	**Power**
1	0.5	25	25	0.193
2	0.5	25	50	0.229
3	0.6	25	25	0.163
4	0.6	25	50	0.190

Figure 4.13.   Summary Table

You can also see that oversampling the control group improves power slightly, 0.229 versus 0.193 for the coefficient of variation of 0.5. However, this is a marginal increase that is probably not worth the added expense.

For the example, use larger sample sizes with equal cell sizes. Select the **Analysis Input** item in the breadcrumb navigation menu at the top of the page.

Then, in the **Sample size** section, change to the first alternate form, **N per group**. Specify sample sizes of 50, 100, 150, and 200, as shown in Figure

4.14.

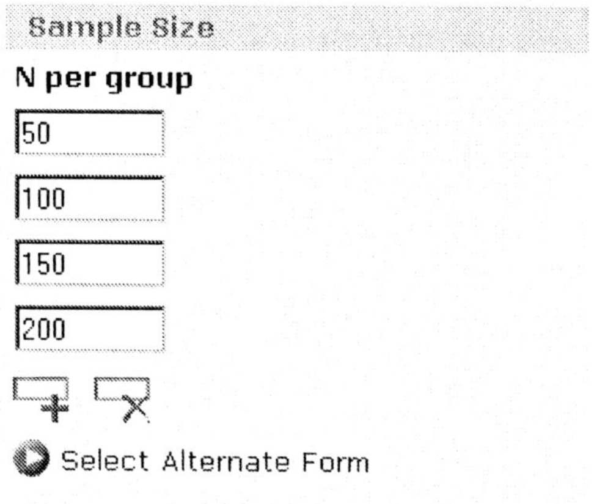

Figure 4.14. Modified Sample Size Values

Table 4.4 contains the modified values of the input parameters for the example.

Table 4.4. Modified Summary of Input Parameters

Parameter	Value
Sample size form	N per group
Sample Size	50, 100, 150, 200

Rerun the analysis by clicking the **OK** button.

Figure 4.15 displays the Summary Table. The largest sample size of 200 (per group) yields a power of 0.72 for a coefficient of variation of 0.5, and 0.599 for one of 0.6. With a total of 400 subjects, you still have a 30% to 40% probability of not detecting the effect even if it exists.

Two-sample t Test for Mean Ratio	

Fixed Scenario Elements	
Distribution	Lognormal
Method	Exact
Number of Sides	1
Alpha	0.05
Geometric Mean Ratio	0.9
Null Geometric Mean Ratio	1

Computed Power			
Index	CV	N Per Group	Power
1	0.5	50	0.296
2	0.5	100	0.471
3	0.5	150	0.611
4	0.5	200	0.720
5	0.6	50	0.242
6	0.6	100	0.380
7	0.6	150	0.499
8	0.6	200	0.599

Figure 4.15. Summary Table for Modified Sample Sizes

Additional Topics

Solving for Sample Size

Several tasks enable you to solve for either total sample size or sample size per group. The sample size per group choice assumes equal group sizes. When solving for total sample size, the group sizes can be equal or unequal. Select the desired result from the drop-down list to the right of the **Calculate** button. The contents of the drop-down list are shown in Figure 4.16.

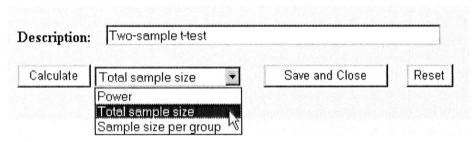

Figure 4.16. Solve for Options in Drop-down List

For either sample size choice, you must specify one or more values for power in the **Power** section. If you frequently use the same value(s) for power, set them as the default on the Preferences page.

If you select total sample size, you must specify whether the group sizes are equal or unequal. Select the appropriate choice in the **Relative Sample Sizes** section. For unequal group sizes, you must specify the relative sample sizes for the two groups. For information on providing relative sample sizes, see the "Using Unequal Group Sizes" section on page 71.

Using Unequal Group Sizes

When solving for either power or total sample size, you may have unequal group sizes. If so, you must provide relative sample sizes for the groups. Weights must be greater than 0 but do not have to sum to 1.

Select the `Total N, Group weights` alternate form in the **Sample Size** section. Enter total sample sizes of 30 and 60 in the **TotalN** table. Select the

Unequal group sizes choice and select the **Enter relative sample sizes** link, as seen in Figure 4.17.

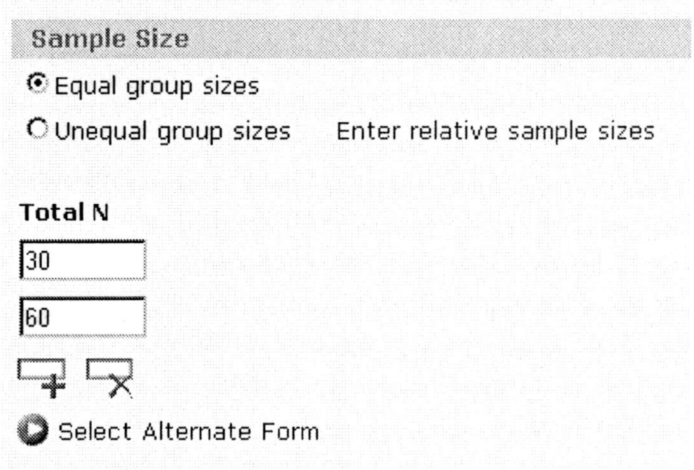

Figure 4.17. Link to Relative Sample Sizes Dialog Page

Figure 4.18 displays the dialog page on which you can enter relative sample sizes. As an example, enter 2 for the first group and 1 for the second. In this case, you are sampling the drug treatment group twice as often as the control group.

The weights control how the total sample size is divided between the two groups. In the example, the sample size for groups 1 and 2 is 20 and 10, respectively, for a total sample size of 30.

Enter a sample size weight for each group.

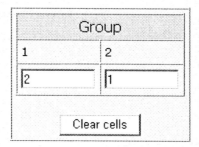

Figure 4.18. Relative Sample Sizes Dialog Page

To clear the values in all cells, press the **Clear cells** button. Press the **OK** button to save the values and return to the Input page.

Chapter 5
Analysis of Variance

Chapter Contents

Chapter 5
Analysis of Variance

Overview

PSS offers power and sample size calculations for analysis of variance in two tasks: One-way ANOVA and General Linear Univariate Models. An optional contrast is available in both tasks.

In the one-way ANOVA task, you can solve for sample size per group as well as total sample size. The contrast for the one-way ANOVA task enables you to select orthogonal polynomials as well as to specify contrast coefficients. For more information about power and sample size analysis for one-way ANOVA, see Chapter 57, "The POWER Procedure" (*SAS/STAT User's Guide*).

In the general linear univariate models task, you specify linear models for a single dependent variable. Type III tests and contrasts of fixed effects are included, and the model can include covariates. For more information about power and sample size analysis for linear univariate models, see Chapter 34, "The GLMPOWER Procedure" (*SAS/STAT User's Guide*).

Example

You are interested in testing how two experimental drugs affect systolic blood pressure relative to a standard drug. You want to include both men and women in the study. You have a two-factor design: a drug factor with three levels and a gender factor with two levels. You choose a main effects only model because you do not expect a drug by gender interaction. You want to calculate the sample size that will produce a power of 0.9 using a significance level of 0.05. You believe that the error standard deviation is between 5 and 7 mm pressure. This is a two-way analysis of variance, so the General Linear Univariate Models task is the appropriate one.

Entering Input Parameters

Start by displaying the Analyses page. Under the **Linear models** section, select `General linear univariate models`. The Input page for the analysis is displayed.

Description

For the example, change the description to `Three blood pressure drugs and gender`.

Solve For

For the example, select `Sample size` from the drop-down list.

Factors

Enter a name for each factor in the design. Factors may contain blanks and other special characters. Do not use an asterisk (*) because a factor name containing it may be confused with an interaction effect. Factor names can be any length, but they must be distinct from one another in the first 32 characters.

Enter names for the two factors, `Drug` and `Gender`, in the factors list and select the appropriate number of levels from the corresponding drop-down list, as shown in Figure 5.1. To add rows to the table for more factors, use the Add a Row (**+**) control below the table.

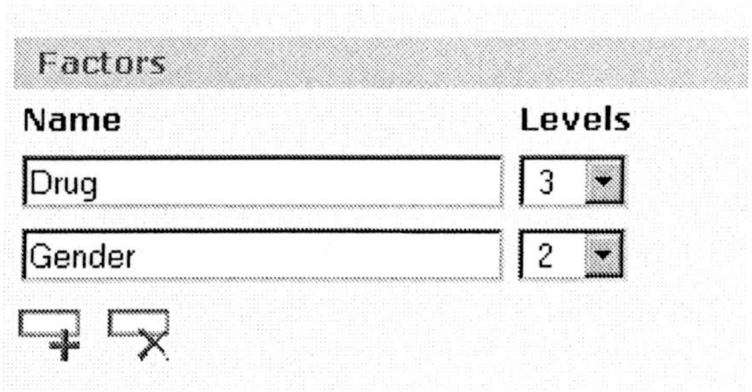

Figure 5.1. Factors and Number of Levels

Model

In the **Model** section you can choose from three models:

`Main effects`
> Only the main effects are included in the model.

`Main effects and interactions`
> The main effects and all possible interactions are included in the model.

`Custom model`
> Selected effects are included in the model. The effects are selected on a dialog page that is accessed by selecting the **Choose model effects** link. For more information on specifying a custom model, see the "Specifying a Custom Model" section on page 92.

For this example, choose the default **Main effects** model, as shown in Figure 5.2.

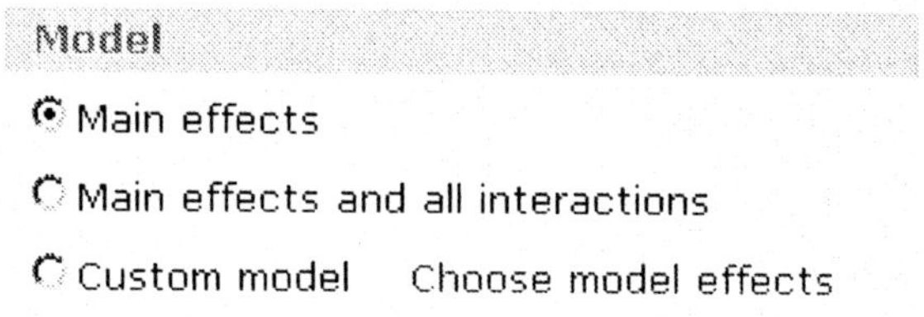

> Model
>
> ⊙ Main effects
>
> ○ Main effects and all interactions
>
> ○ Custom model   Choose model effects

Figure 5.2.   Model with Main Effects Selected

Alpha

Alpha is the significance level that is the probability of falsely rejecting the null hypothesis. If you frequently use the same value(s) for alpha, set them as the default on the Preferences page.

Specify a single significance level of 0.05 in the **Alpha** section.

Means

You must enter projected cell means for each cell of the design. Cell means are entered on a dialog page. To display the cell means dialog, select the **Enter means** link in the **Means** section, as is shown in Figure 5.3.

Means

Enter means

Figure 5.3. The Enter Means Link to Access Means Dialog Page

The cell means for the example are listed in Table 5.1.

Table 5.1. Cell Means

Gender	Drug		
	Experimental 1	**Experimental 2**	**Standard**
Males	130	128	125
Females	125	121	118

The completed Enter Means dialog page is shown in Figure 5.4.

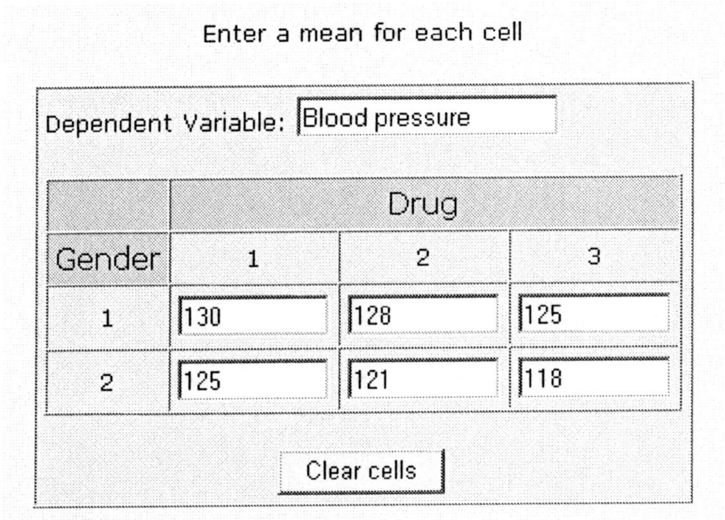

Enter a mean for each cell

Dependent Variable: Blood pressure

Gender	Drug		
	1	2	3
1	130	128	125
2	125	121	118

Clear cells

Figure 5.4. Means Dialog Page

You can also specify the name of the dependent variable; in this example, `Blood pressure` is used. After entering the means, press the **OK** button to save the values and return to the previous window.

Standard Deviation

Specify one or more conjectured error standard deviations in the **Standard Deviation** section. This is the same as the root mean square error.

For this example, enter two standard deviations, 5 and 7, as shown in Figure 5.5.

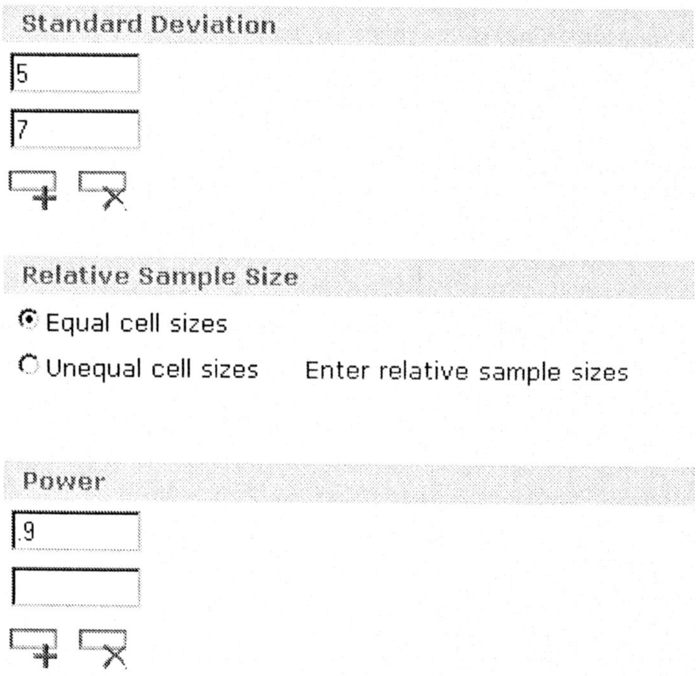

Figure 5.5. Standard Deviations, Relative Sample Sizes, and Powers

Relative Sample Size

When solving for sample size, it is necessary to specify whether the cell sample sizes are equal or unequal. If cell sizes are unequal, relative sample size weights must also be specified. For more information on providing sample size weights, see the "Using Unequal Cell Sizes" section on page 90.

For the example, select `Equal cell sizes` in the **Relative Sample Size** section, as shown in Figure 5.5.

Power

Specify one or more powers in the **Power** section.

For this example, enter a single power of 0.9, as shown in Figure 5.5.

Summary of Input Parameters

Table 5.2 contains the values of the input parameters for the example.

Table 5.2.   Summary of Input Parameters

Parameter	Value
Model	Main effects
Alpha	0.05
Means	See Table 5.1
Standard Deviation	5, 7
Relative sample sizes	Equal cell sizes
Power	0.9

Results Options

Select all three results: the `Summary table of parameters`, the `Power by Sample Size` graph, and `Narratives of selected results`. The graph consists of four points, one for each of the four scenarios created by combining the two factor main effects with the two standard deviations. This graph will not be very informative, so specify a range of powers to the vertical power axis.

Select the **Customize** link beside the graph choice in the **Graph** section. The Customize Graph dialog page is displayed. In the Powers table of the **Value Ranges** section, enter a minimum power of 0.75 and a maximum power of 0.95, as shown in Figure 5.6. Press the **OK** button to close the dialog page.

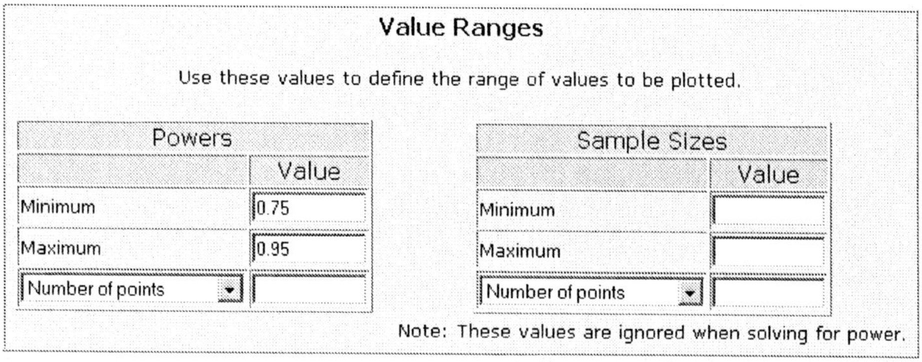

Figure 5.6. Value Ranges on Customize Graph Dialog Page

Now, back on the Input page, press the **Calculate** button to perform the analysis.

Viewing Results

In the Computed N Total table on the Results page (Figure 5.7), sample sizes are listed for each combination of factor and standard deviation. You will need a total sample size between 60 and 108 to yield a power of 0.9 for the Drug effect if the standard deviation is between 5 and 7. You will need a sample size of half that for the Gender effect.

General linear univariate models

Fixed Scenario Elements	
Dependent Variable	Blood pressure
Alpha	0.05
Nominal Power	0.9

Computed N Total						
Index	Source	Std Dev	Test DF	Error DF	Actual Power	N Total
1	Drug	5	2	56	0.921	60
2	Drug	7	2	104	0.905	108
3	Gender	5	1	26	0.916	30
4	Gender	7	1	50	0.903	54

Figure 5.7. Summary Table

The power by sample size graph appears in Figure 5.8. One curve is displayed for each standard deviation and factor combination. From the upper portions of the curves, you can see the range of sample sizes that correspond to a power of 0.9 or greater.

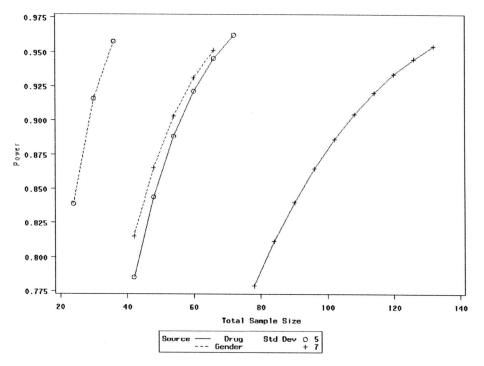

Figure 5.8. Power by Sample Size Graph

The narrative for the first scenario, the Drug effect and the standard deviation of 5, is created by default. Note that the cell means are not included in the narrative description:

> For the usual F test of the Drug effect in the general linear uni-variate model with fixed class effects [Blood pressure = Drug Gender] using a significance level of 0.05, assuming the specified cell means and an error standard deviation of 5, a total sample size of 60 assuming a balanced design is required to obtain a power of at least 0.9. The actual power is 0.921.

For information on creating narratives for other scenarios, see the "Creating Narratives" section on page 44.

Additional Topics

Adding a Contrast

Contrasts are optional. A contrast can be added when using a main effects model. The contrast coefficients are entered on a dialog page. To access the dialog page, select the **Enter contrast** link in the **Contrast** section, as shown in Figure 5.9.

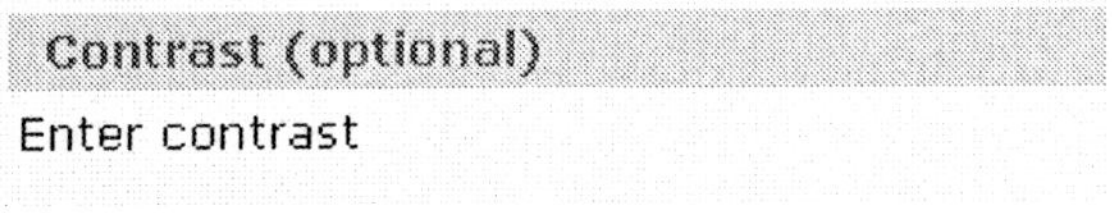

| Contrast (optional) |
| Enter contrast |

Figure 5.9.　Enter Contrast Link to Access Contrast Dialog Page

Use the **Label** field to provide a label for the contrast. It should be different from all of the factor names and all interactions in the model.

Enter a valid contrast by specifying at least two coefficients per factor for at least one factor. As noted, it is not necessary to enter zeros; blanks are considered to be zeros.

To clear all of the contrast coefficients, press the **Clear coefficients** button. To remove a previously defined contrast, clear the coefficients and then press the **OK** button.

In this example you are interested in comparing the two experimental drugs to the standard drug. As shown in Figure 5.10, the contrast coefficients are 0.5, 0.5, and −1 for the three levels of the Drug effect. Press the **OK** button to save the values and return to the Input page.

Enter contrast coefficients

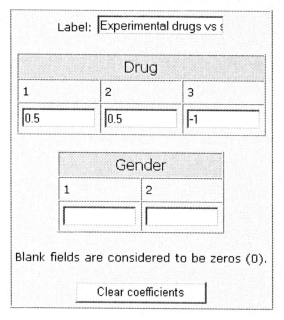

Figure 5.10. Contrast Dialog Page with Coefficients

Figure 5.11 shows the two scenarios for the contrast at the bottom of the Computed N Total table. The two scenarios also appear in the graph but are not shown here.

			Computed N Total				
Index	Type	Source	Std Dev	Test DF	Error DF	Actual Power	N Total
1	Effect	Drug	5	2	56	0.921	60
2	Effect	Drug	7	2	104	0.905	108
3	Effect	Gender	5	1	26	0.916	30
4	Effect	Gender	7	1	50	0.903	54
5	Contrast	Experimental drugs vs standard	5	1	62	0.924	66
6	Contrast	Experimental drugs vs standard	7	1	116	0.909	120

Figure 5.11. Computed N Total Table for the Contrast

Solving for Power

In addition to solving for sample size, you can also solve for power. Select the desired result from the drop-down list to the right of the **Calculate** button. Figure 5.12 shows the two choices in the drop-down list.

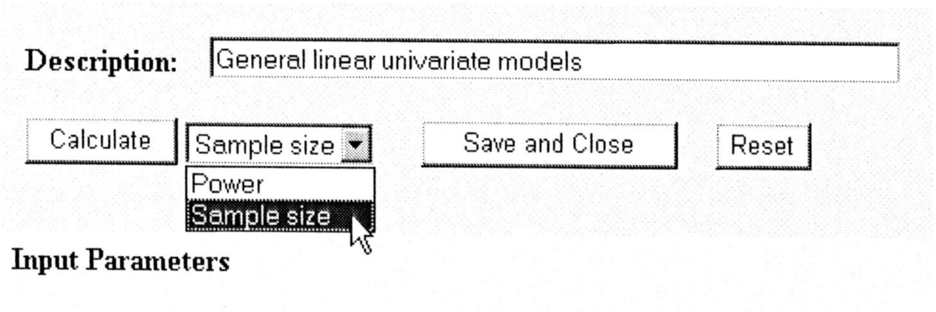

Figure 5.12.   Contents of Solve For Drop-down List

When solving for power, you must provide sample size information. For the General Linear Univariates Model analysis, you provide this information by using one of two alternate forms. To choose the desired alternate form, select the **Select Alternate Form** link in the **Sample Size** section, then select the desired form from the list. The alternate forms are:

`N per cell`

> Enter the sample size for the cells. Cell sizes are assumed to be equal. Sample size is reported in the summary results table as total sample size.

`Total N, Cell weights`

> Enter the total sample size and specify whether cell sizes are to be equal or unequal. Choose `Equal cell sizes` or `Unequal cell sizes` in the **Sample Size** section. For unequal cell sizes, you also enter cell weights. Select the **Enter relative sample sizes** link to display a dialog page that is used to enter the data. For more information on using unequal cell sizes, see the "Using Unequal Cell Sizes" section on page 90.

Using Unequal Cell Sizes

If you may have unequal cell sizes, you must enter relative sample size weights for the cells. Weights do not have to sum to 1 across the cells. Some weights can be zero but enough weights must be greater than zero so that the effects and contrasts are estimable.

Select the `Total N, Cell weights` alternate form in the **Sample Size** section. Select the `Unequal cell sizes` choice and select the **Enter relative sample sizes** link, as seen in Figure 5.13.

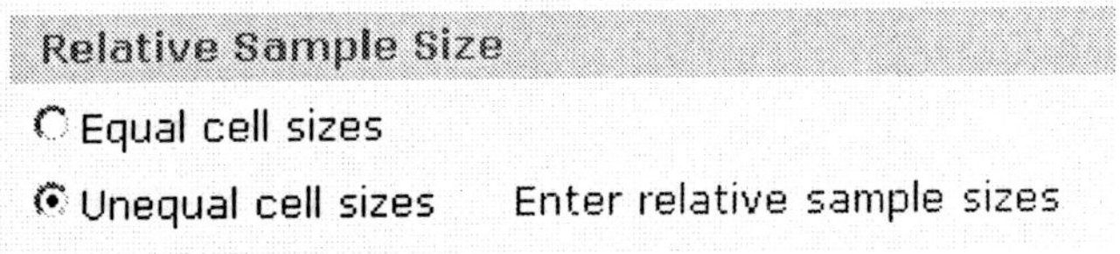

Relative Sample Size

◯ Equal cell sizes

◉ Unequal cell sizes   Enter relative sample sizes

Figure 5.13.   Link to Relative Sample Sizes Dialog Page

Figure 5.14 shows a dialog page on which you can enter relative sample sizes. As an example, enter the sample size weights from Table 5.3.

Table 5.3.   Sample Size Weights

Gender	Drug		
	Experimental 1	**Experimental 2**	**Standard**
Males	1	1	2
Females	1	1	2

In this case, you want the sample size of the standard group to be twice that of each of the two experimental groups. Press the **OK** button to save the values and return to the Input page.

Enter a sample size weight for each cell.

Gender	Drug		
	1	2	3
1	1	1	2
2	1	1	2

Clear cells

Figure 5.14. Relative Sample Sizes Dialog Page

Figure 5.15 shows the summary table for the Drug by Gender example.

Fixed Scenario Elements	
Dependent Variable	Blood pressure
Weight Variable	_Weight_
Alpha	0.05
Nominal Power	0.9

	Computed N Total					
Index	Source	Std Dev	Test DF	Error DF	Actual Power	N Total
1	Drug	5	2	52	0.910	56
2	Drug	7	2	100	0.902	104
3	Gender	5	1	28	0.944	32
4	Gender	7	1	52	0.926	56

Figure 5.15. Summary Table for Unbalanced Design Example

Specifying a Custom Model

In the **Model** section you can choose from three types of models: a `Main effects` model, a `Main effects and all interactions` model, and a `Custom model`. To specify a custom model, select the `Custom model` choice and, then, select the **Choose model effects** link, as shown in Figure 5.16.

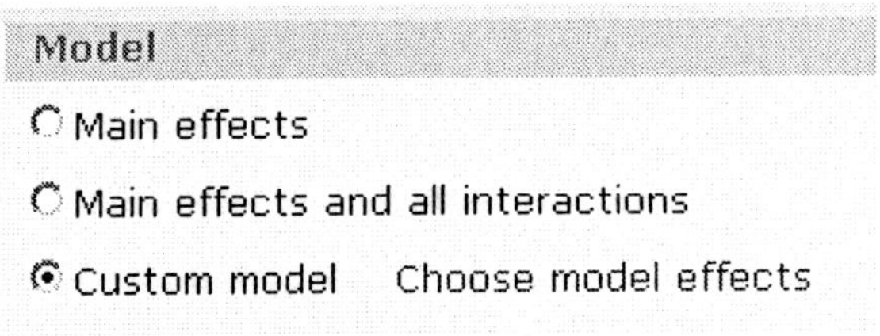

Figure 5.16. Link to Custom Model Dialog Page

A dialog page is displayed that contains a list of all model main effects and interactions. Select the effects that you want included in the model. The example shown in Figure 5.17 has three main effects and all possible interactions. The figure shows how you can choose certain effects to be in the model.

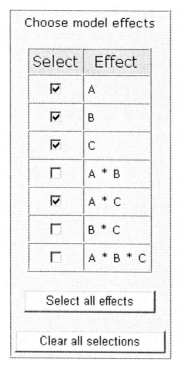

Figure 5.17. Custom Model Dialog Page

Press the **OK** button to save the information and return to the previous page.

Including Covariates

Covariates are optional. If you have covariates, include the total number of degrees of freedom for all covariates. Add the number of continuous covariates and the sum of the degrees of freedom of the classification covariates and select this total in the **Number of Covariates** drop-down list. For example, with two continuous covariates and a single classification covariate factor with three levels, the total would be $2 + (3 - 1) = 4$.

Also, you must enter the correlation between the dependent variable and the set of covariates. Two alternate forms are available: the multiple correlation and the proportional reduction in variance. Select the desired form and enter one or more values.

The multiple correlation is between the set of covariates and the dependent variable. Proportional reduction in variation is how much the variance of the dependent variable is reduced by the inclusion of the covariates, expressed as a proportion between 0 and 1.

Figure 5.18 illustrates four covariates and a proportional reduction in variation of 0.3. The results for the analysis are not shown.

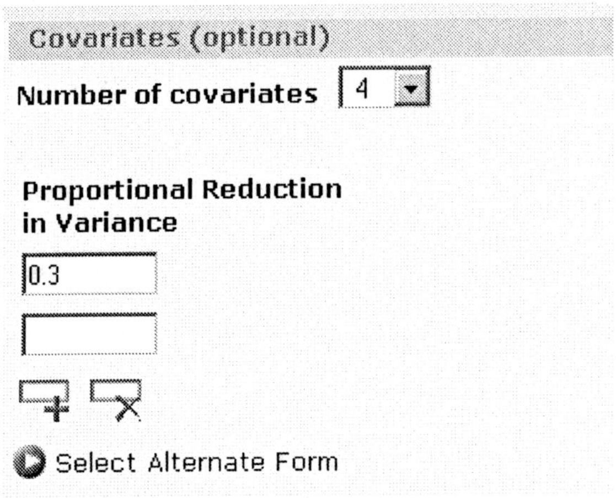

Figure 5.18. Covariates with Proportional Reduction in Variance Form

Chapter 6
Two-Sample Survival Rank Tests

Chapter Contents

Chapter 6
Two-Sample Survival Rank Tests

Overview

Survival analysis often involves the comparison of survival curves. PSS provides sample size and power calculations for two-sample survival rank analyses. Several rank tests are available: Gehan, log-rank, and Tarone-Ware as well as several ways of specifying the survival functions. For more information about power and sample size analysis for survival rank tests, see Chapter 57, "The POWER Procedure" (*SAS/STAT User's Guide*).

Example

You want to compare survival rates for an existing and a new cancer treatment. You intend to use a log-rank test to compare the overall survival curves for the two treatments. You want to determine a sample size to achieve a power of 0.8 for a two-sided test using a balanced design, with a significance level of 0.05.

The survival curve of patients for the existing treatment is known to be approximately exponential with a median survival time of five years. You think that the proposed treatment will yield a survival curve described by the times and probabilities listed in Table 6.1. Patients are to be accrued uniformly over two years and followed for three years.

Table 6.1. Survival Probabilities for Proposed Treatment

Time	Probability
1	0.95
2	0.90
3	0.75
4	0.70
5	0.60

Start by displaying the Analyses page. Under the **Survival Analysis** section, select `Two-sample survival rank tests`. The Input page for the analysis is displayed.

Entering Input Parameters

Description

For the example, change the description to `Comparing cancer treatments using two-sample survival rank test`.

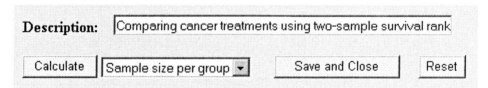

Description: | Comparing cancer treatments using two-sample survival rank |

Calculate | Sample size per group ▾ | Save and Close | Reset

Figure 6.1. Description and Solving For Sample Size

Solve For

For this analysis, you can choose to solve for power, total sample size, or sample size per group. Select the desired quantity from the drop-down list to the right of the **Calculate** button.

For this example, select `Sample size per group` from the drop-down list, as shown in Figure 6.1. For information on calculating total sample size, see the "Solving for Sample Size" section on page 71.

Test

Several rank tests are available: Gehan, log-rank, and Tarone-Ware. The Gehan test is most sensitive to survival differences near the beginning of the study period, the log-rank test is uniformly sensitive throughout the study period, and the Tarone-Ware test is somewhere in between.

For this example select the `Log-rank` choice in the **Test** section, as shown in Figure 6.2.

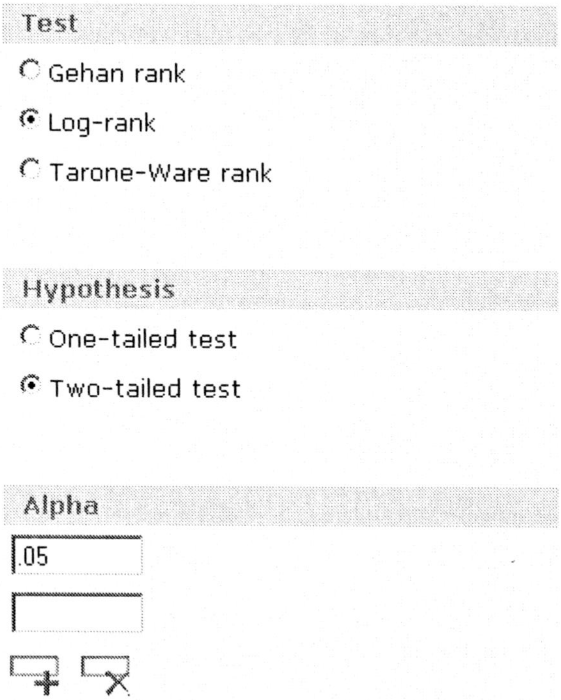

Figure 6.2. Test, Hypothesis, and Alpha Options

Hypothesis

You can choose either a one- or two-tailed test. For the one-tailed test, the alternative hypothesis is assumed to be in the same direction as the effect. If you do not know the direction of the effect, that is, whether it is positive or negative, the two-tailed test is appropriate. If you know the effect's direction, the one-tailed test is appropriate. If you specify a one-tailed test and the effect is in the unexpected direction, the results of the analysis are invalid.

Select the `Two-tailed test` choice in the **Hypothesis** section, as shown in Figure 6.2.

Alpha

Alpha is the significance level that is the probability of falsely rejecting the null hypothesis. If you frequently use the same value(s) for alpha, set them as the default on the Preferences page.

Enter the desired significance level of 0.05 into the first cell of the Alpha table if it is not already the default value, as shown in Figure 6.2.

Survival Functions

In the **Survival functions** section, there are four alternate forms for entering survival functions. The first three apply only to exponential curves; the fourth applies to both piecewise linear and exponential curves.

`Group median survival times`
> Enter median survival times for the two groups.

`Group hazards`
> Enter hazards for the two groups.

`Group 1 hazards, Hazard ratios`
> Enter hazards for group 1 and hazard ratios.

`Survival curves`
> Enter survival probabilities and their associated times for each of several curves. Specify the number of curves; at least two curves are required. Then, select the **Enter survival probabilities** link to display a dialog page that is used to enter the survival times and probabilities.

Examine the alternatives by selecting the **Select Alternate Form** link. This example uses the fourth form. For information on using the other forms, see the "Using the Other Survival Curve Forms" section on page 111.

For the example, select the fourth form, `Survival curves`, as shown in Figure 6.3. Select the value, 2, from the **Number of survival curves** drop-down list. Select the **Enter survival probabilities** link. The Survival Curves dialog page is displayed.

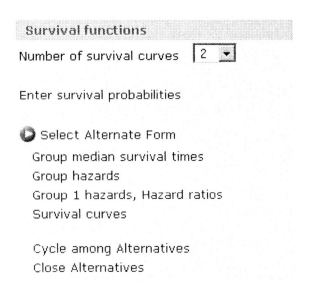

Figure 6.3. Survival Functions Section with Number of Curves

You provide two sets of information on this dialog page. Use the upper panel to assign times and survival probabilities for each survival curve. Use the lower panel to assign each curve to one of two groups. You can also provide a label for each curve. The labels should be unique.

When entering probabilities in the upper panel, enter a value only when the probability for a survival curve changes. For example, if the probabilities for curve 1 at time 1 and 2 is 0.9 and at time 3 is 0.8, only enter 0.9 for time 1 and 0.8 for time 3. Because survival probabilities may change at different times for the two (or more) curves, you will not necessarily enter a probability in each row for each curve. Such a table is shown in Figure 6.4.

Enter survival probabilities for each curve.

Survival probabilities

Curve

Time	1	2
8	.95	
10		.92
15	.87	
19		.88
21	.80	
25		.80
28	.72	
30	.70	.75

Number of survival times   8 ▼

Figure 6.4. Example Table of Survival Times and Probabilities

To enter an exponential survival curve, enter a single probability. In the example, the exponential curve for the existing treatment is defined by a probability of 0.5 at time 5.

The units of time for the survival curves must correspond to the units for the accrual, follow-up, and total times.

You can add or delete table rows in two ways. Select a value from the **Number of survival times** drop-down list. This sets the number of rows in the table. You can either increase or decrease the number of rows. You can also add or delete one table row at a time by pressing the Add a Row (+) or Delete Last Row (−) control, respectively, These controls are displayed beside the **Number of survival times** drop-down list.

Enter the following values. In the upper panel, enter the values 1 through 5 in the first column as the times in the first 5 rows. For the first curve (the second column), enter the value 0.5 for time 5, leaving the other rows blank. For the second curve (the third column), enter the values 0.95, 0.9, 0.75, 0.7, and 0.6 in the five rows, respectively. Figure 6.5 illustrates the table with these values.

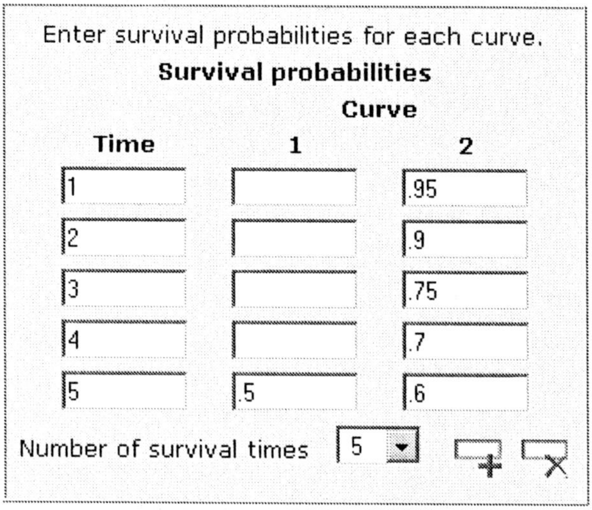

Enter survival probabilities for each curve.

Survival probabilities

Curve

Time	1	2
1		.95
2		.9
3		.75
4		.7
5	.5	.6

Number of survival times 5

Figure 6.5.   Survival Times and Probabilities

In the lower panel, change the label of Curve 1 to **Existing treatment** and of Curve 2 to **Proposed treatment**. Select the **Group 1** choice for the existing treatment curve and the **Group 2** choice for the proposed treatment curve. These group selections may be the default ones. Figure 6.6 shows the resulting values.

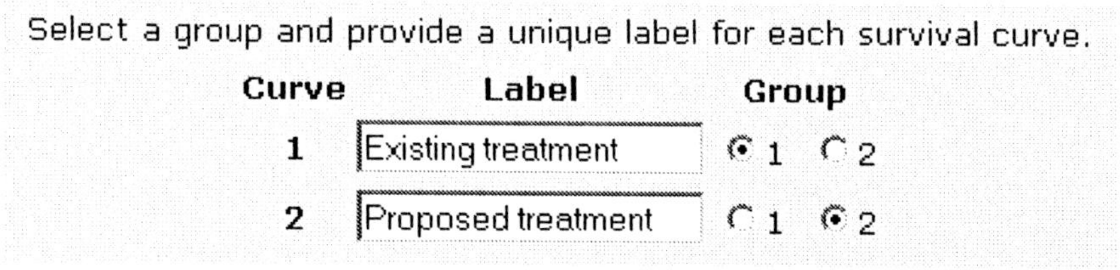

Select a group and provide a unique label for each survival curve.

Curve	Label	Group
1	Existing treatment	⊙ 1 ○ 2
2	Proposed treatment	○ 1 ⊙ 2

Figure 6.6.   Survival Curves with Labels and Group Assignments

Press the **OK** button to save the values and settings and return to the previous page.

You can also compare several survival curves. For example, if you have two scenarios, A and B, for group 1's curve and two scenarios, C and D, for group 2's curve, then specify probabilities for the four curves and assign A and B to group 1 and C and D to group 2.

Accrual and Follow-up Times

Accrual time is the period during which subjects are brought into the study. Follow-up time is the period during which subjects are observed after all subjects have been included in the study. Total time is the sum of accrual and follow-up time. The units of time for the accrual, follow-up, and total times must correspond to the units used when specifying the survival curves.

When entering survival curves, the sum of the accrual and follow-up times must be less than the largest time for each survival curve. This does not apply to survival curves represented by a single time, which represent exponential curves.

In the **Accrual and follow-up times** section, there are three alternate forms for entering accrual and follow-up times:

`Accrual, follow-up times`
> Enter accrual and follow-up times.

`Accrual, total times`
> Enter accrual and total times.

`Follow-up, total times`
> Enter follow-up and total times.

Examine the alternatives by selecting the **Select Alternate Form** link.

For the example, select the first form, `Accrual and follow-up time`. Then enter a single value of 2 in the Accrual table and a value of 3 in the Follow-up table, as shown in Figure 6.7.

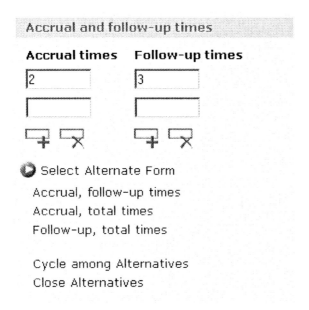

Figure 6.7. Accrual and Follow-up Times

Power

Specify one or more powers to use when calculating sample size.

For the example, enter a single value of 0.8.

Summary of Input Parameters

Table 6.2 contains the values of the input parameters for the example. Table 6.3 contains times and probabilities for the two survival curves.

Table 6.2. Summary of Input Parameters

Parameter	Value
Test	Log-rank
Hypothesis	Two-tailed test
Alpha	0.05
Survival Function form	Survival curves
Survival curves	See Table 6.3
Accrual and follow-up times form	Accrual and follow-up
Accrual times	2
Follow-up times	3
Power	0.8

Table 6.3. Survival Times and Probabilities

	Probabilities	
Time	Existing treatment	Proposed treatment
1		0.95
2		0.90
3		0.75
4		0.70
5	0.5	0.60

Results Options

Request all three results by selecting the check boxes for the `Summary table of parameters`, the `Power by Sample Size` graph, and `Narratives for selected results`.

Specifying only one power as in this example produces a graph with a single point. You may be interested in a plot of sample sizes for a range of powers, say, between 0.75 and 0.85. You do so by customizing the graph by specifying the values for the power axis. Also, you may want to change the appearance of the graph to have sample size (per group) on the vertical axis and power on the horizontal axis.

Select the **Customize** link for the Power by Sample Size graph in the **Graph** section. The Customize Graph dialog page is displayed, as shown in Figure 6.8.

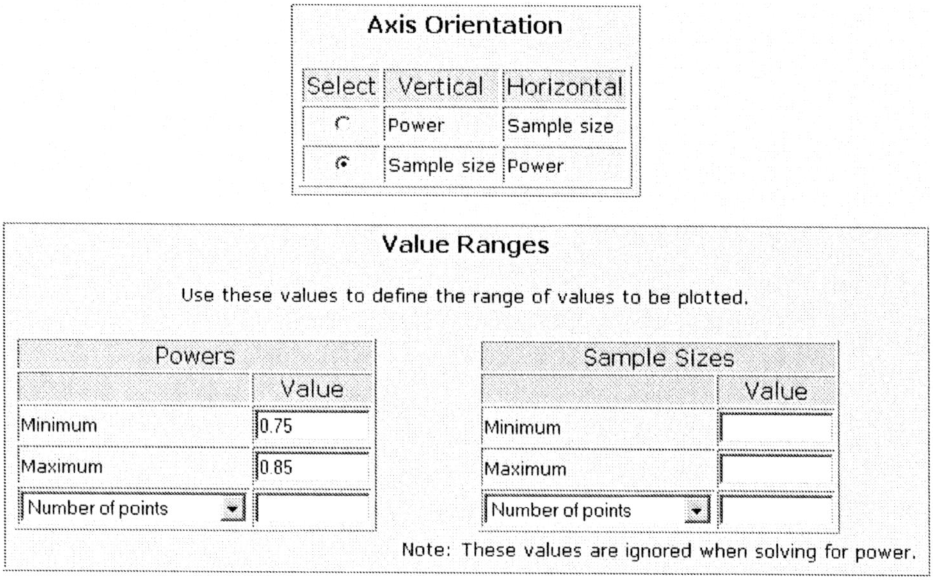

Axis Orientation

Select	Vertical	Horizontal
○	Power	Sample size
⊙	Sample size	Power

Value Ranges

Use these values to define the range of values to be plotted.

Powers	Value
Minimum	0.75
Maximum	0.85
Number of points ▼	

Sample Sizes	Value
Minimum	
Maximum	
Number of points ▼	

Note: These values are ignored when solving for power.

Figure 6.8. Customize Graph Dialog Page

For the **Axis Orientation** section, select the second choice – `Sample size` on the Vertical axis and `Power` on the Horizontal axis, as shown in the top panel of Figure 6.8.

For the **Value Ranges** section, in the Power table enter a minimum of 0.75 and a maximum of 0.85. This sets the range of values on the axis for powers. The completed Value Ranges section of the dialog is displayed in the bottom panel of Figure 6.8.

Press the **OK** button to save the values that you have entered and return to the previous page.

On the Input page, press the **Calculate button** to perform the analysis. If there are no errors in the input parameter values, the Results page appears. If there are errors in the input parameter values, you will be prompted to correct them.

Viewing Results

The results are listed in the table of contents on the left of the page. Select each result that you want to view.

Summary Table

As shown in Figure 6.9, the **Fixed Scenario Elements** and **Computed N Per Group** tables include the values of the input parameters and the computed quantity (in this case, sample size per group, **N per group**). The sample size per group for the single requested scenario is 226.

Log-Rank Test for Two Survival Curves

Fixed Scenario Elements	
Method	Lakatos normal approximation
Number of Sides	2
Accrual Time	2
Follow-up Time	3
Alpha	0.05
Group 1 Survival Curve	Existing treatment
Form of Survival Curve 1	Exponential
Group 2 Survival Curve	Proposed treatment
Form of Survival Curve 2	Piecewise Linear
Nominal Power	0.8
Number of Time Sub-Intervals	12
Group 1 Loss Exponential Hazard	0
Group 2 Loss Exponential Hazard	0

Computed N Per Group	
Actual Power	N Per Group
0.800	226

Figure 6.9.   Summary Table

Power by Sample Size Graph

As you can see in Figure 6.10, the graph is slightly curved upward with larger powers associated with larger sample sizes. Sample size is on the vertical axis as requested on the Customize Graph dialog page.

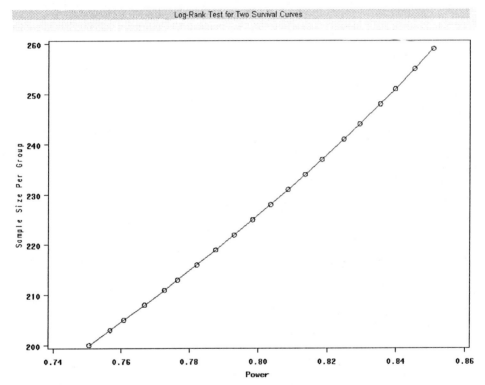

Figure 6.10.   Power by Sample Size Graph

Narratives

The narrative for this task does not include the survival times and probabilities for the survival curves:

> For a log-rank test comparing two survival curves with a two-sided significance level of 0.05, assuming uniform accrual with an accrual time of 2 and a follow-up time of 3, a sample size of 226 per group is required to obtain a power of at least 0.8 for

the exponential curve, "Existing treatment," and the piecewise linear curve, "Proposed treatment." The actual power is 0.800.

For information about selecting additional narratives when multiple scenarios are present, see the "Creating Narratives" section on page 44.

Additional Topics

Using the Other Survival Curve Forms

Survival functions can be specified as median survival times, hazards, or a combination of hazards for one group and hazard ratios. These all assume exponential curves.

Say that you are interested in comparing the proposed and existing treatments using their median survival times. The survival times are five years and four years for the two groups, respectively.

As described in the the "Survival Functions" section on page 100, examine the list of alternate forms for the **Survival functions** section of the Input page and select the **Median survival times** choice.

You can enter one or more sets of two median survival times. For the example enter 5 and 4 in the first row of the table. The completed table is shown in Figure 6.11. The results of the analysis are not shown.

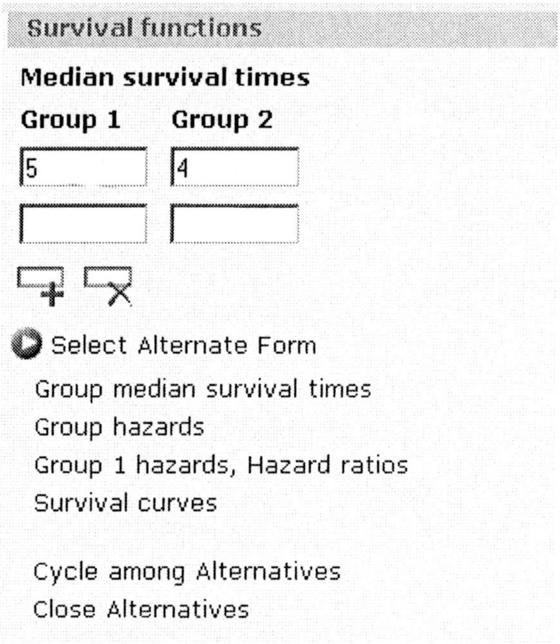

Figure 6.11. Median Survival Times and List of Alternate Forms

Appendix A
Web Applications

Appendix Contents

Appendix A
Web Applications

Web Application

The SAS/STAT Power and Sample Size (PSS) application is a Web application. It requires Web server software to run and is accessed using the Microsoft Internet Explorer Web browser.

One advantage of a Web application is that many users can access the application that is running on one machine by means of a company's intranet. Each user accesses the application from a Web browser that is running on his or her own machine.

PSS is installed separately from the SAS/STAT product. It is included on the SAS Mid-Tier Components CD. See the installation instructions for the Mid-Tier Components CD in the installation package.

The application is configured at installation time to run in one of two configurations – local or remote.

In the local configuration, the browser, the Web server, the application, and SAS 9.1 are all installed on one machine. Access to the Internet is not necessary when using the application in its local configuration.

In the remote configuration, the browser, SAS 9.1, and the Web server and application may be installed on different machines. Access to the Internet is necessary when using the application in its remote configuration.

Accessing the Application

To access the application, you must provide the appropriate address to your Web browser. The address is known as a uniform resource locator (URL). Obtain it from your Web administrator.

The URL is composed of the server name, the server port number, the name of the application, and the starting point for the application, `Power`. The application name and default port number varies with different Web servers. The form of the URL is

http://servername:portnumber/applicationname/Power

For a server named "ourserver," a port number of 8080, and an application name of "pss," the URL might be

http://ourserver:8080/pss/Power

When using PSS in the local configuration, a special server name of "local-host" can be used:

http://localhost:8080/pss/Power

Check with your network administrator or consult the documentation for the Web server for the correct syntax of the URL.

Important Tip

Do not use the **Back** and **Forward** buttons on your Web browser to move from page to page within the application. Doing so bypasses the application and produces unpredictable results. Instead, use the navigation bar, bread-crumb trail, or buttons on each Web page.

Timeouts

If you do not use the application for a period of about 30 minutes, any information that you have entered and not saved may be lost.

Web servers respond only to requests from a browser; they cannot send information to a browser in the absence of a request. Therefore, a Web server has no way of determining if your session is still active. Put another way, if a Web server has not received a request from your browser recently, it doesn't know if you are still working on the application or have moved onto other work or have closed the browser.

The Web server stores information about your recent use of the application. If the Web server has not heard from you in a specified amount of time (usually 30 minutes), it discards the information. You must send a request to the server in order for this timeout period to be reset. A request is made of the Web server when you request another page by using the application's navigation bar or an action button that displays another page.

If a timeout occurs in the application, the application displays the Projects page along with a message that a timeout condition has occurred. If you think you will not be using the application for a while and you have entered information that you want saved, be sure to save the data by using the **Save and Close** button on the Input page (see the "Actions" section on page 37).

Index

Your Turn

If you have comments or suggestions about *Getting Started with the SAS®
Power and Sample Size Application,* please send them to us on a photocopy of
this page or send us electronic mail.

For comments about this book, please return the photocopy to

SAS Publishing
SAS Campus Drive
Cary, NC 27513
E-mail: **yourturn@sas.com**

For suggestions about the software, please return the photocopy to

SAS Institute Inc.
Technical Support Division
SAS Campus Drive
Cary, NC 27513
E-mail: **suggest@sas.com**

Printed in the United States
25797LVS00002B/1-2